An atlas of edge-reversal dynamics

Submission of proposals for consideration

Suggestions for publication, in the form of outlines and representative samples, are invited by the Editorial Board for assessment. Intending authors should approach one of the main editors or another member of the Editorial Board, citing the relevant AMS subject classifications. Alternatively, outlines may be sent directly to the publisher's offices. Refereeing is by members of the board and other mathematical authorities in the topic concerned, throughout the world.

Preparation of accepted manuscripts

On acceptance of a proposal, the publisher will supply full instructions for the preparation of manuscripts in a form suitable for direct photo-lithographic reproduction. Specially printed grid sheets can be provided. Word processor output, subject to the publisher's approval, is also acceptable.

Illustrations should be prepared by the authors, ready for direct reproduction without further improvement. The use of hand-drawn symbols should be avoided wherever possible, in order to obtain maximum clarity of the text.

The publisher will be pleased to give guidance necessary during the preparation of a typescript and will be happy to answer any queries.

Important note

In order to avoid later retyping, intending authors are strongly urged not to begin final preparation of a typescript before receiving the publisher's guidelines. In this way we hope to preserve the uniform appearance of the series.

CRC Press UK

Chapman & Hall/CRC Statistics and Mathematics
Pocock House
235 Southwark Bridge Road
London SE1 6LY
Tel: 020 7450 7335

Valmir C Barbosa

An atlas of edge-reversal dynamics

CRC Press
Taylor & Francis Group
Boca Raton London New York

CRC Press is an imprint of the
Taylor & Francis Group, an **informa** business

Library of Congress Cataloging-in-Publication Data

Barbosa, Valmir C.
    An atlas of edge-reversal dynamics/by Valmir C. Barbosa.
        p. cm. -- (Chapman & Hall/CRC research notes in mathematics series)
    Includes bibliographical references and index.
    ISBN 1-58488-209-3 (alk. paper)
    1. Electronic data processing--Distributed processing. 2. Parallel processing (Electronic computers) I. Title. II. Series.

QA76.9.D5 B355 2000
004'.36--dc21   00-043020

© 2001 by Chapman & Hall/CRC

No claim to original U.S. Government works
International Standard Book Number 1-58488-209-3
Library of Congress Card Number 00-043020

This book is dedicated to
Alzira, Leonardo, Julia, and Isabel.

Contents

Preface

This book is a reference volume on the dynamics of scheduling by edge
reversal (SER). SER is a distributed mechanism for scheduling agents in
a computer system to perform tasks in parallel, subject to the constraint
that no parallelism is to exist within an agent's neighborhood. Depending
on how one defines an agent and its neighborhood, SER has important
applications in computer science, especially in resource-sharing systems and
within artificial intelligence and optimization.

SER is based on evolving acyclic orientations of an undirected graph.
Starting at any initial acyclic orientation, an infinite sequence of acyclic
orientations is obtained by effecting a very simple transformation on the
current orientation to obtain the next. This transformation is the reversal
of the orientations of all edges adjacent to sinks (nodes in the graph whose
adjacent edges are all oriented inward). As the graph's orientations evolve,
so do the sets of sinks, forming the basis for the application of SER as an
agent scheduler.

In this book, I have adopted the view of SER as a dynamical system whose state space is the set of acyclic orientations of an undirected
graph. From the perspective of such dynamics, the coarse structure of
this set of orientations comprises several basins of attraction, each one en-
compassing some of the graph's acyclic orientations. The performance of
SER in terms of how much parallelism it is ultimately capable of providing
when scheduling agents is determined by the basin of attraction to which
its operation is confined. Choosing appropriate initial conditions is then

the most central problem to be solved in connection with the dynamics of SER.

The measure of parallelism that has to be optimized when solving this problem is akin to several of the underlying graph's chromatic indicators, and is, like most of those, of probably intractable optimization in general. This book is an attempt to subsidize the search for good heuristics to optimize the performance of SER, by providing data on the mechanism's behavior over as comprehensive a set of situations as can be presented in a single volume.

The book comprises two parts. Part I is a review of the origins of SER, its properties, and its applications. It also contains data from statistics and correlations that have been computed over several graph classes of interest.

Part II is by far the largest of the two parts, comprising the atlas proper. This part of the book contains graphical representations of all basins of attraction for all graphs in selected classes. Naturally, the possible choices as to which classes would go into the book were overwhelmingly too many, and each one quite vast. My final decision arose not only from space constraints, but mostly from the richness of information that the data on the selected classes would be able to convey.

Being a repository of results and data on SER, I have in general opted to not include in the book proofs of the results asserted. I have, however, attempted to provide a comprehensive set of references where the missing proofs can be looked up.

Finally, in addition to the great applicability of SER that I already mentioned, there has been another major motivation driving me to undertake the preparation of a volume like this one. As with many other similarly fascinating subjects, the study of SER reveals very clearly the emergence of complex dynamic behavior from very simple transition rules. SER does, as such, provide the opportunity for the study of complex graph dynamics with strong motivation from several application areas.

Notation

The notation I use is reasonably standard, with the possible exception of \ to denote set difference.

Acknowledgments

Work on this book has taken place at two institutions. First at the Federal University of Rio de Janeiro, where most of the research and early writing were done. Then at the Computer Science Division of the University of California, Berkeley, where the writing was finished. At these two places, and elsewhere, I wish to acknowledge several people who, during the book's lengthy preparation, contributed in various ways to its success-ful completion. These include true SER enthusiasts, people with whom I discussed SER-related material, and colleagues who contributed with the means for the book to be completed. They are Mario Benevides, Charlie Colbourn, Luerbio Faria, Luís Favre, Felipe França, Brendan McKay, Jayme Szwarcfiter, and Lotfi Zadeh.

Partial support for my research has been provided, over the years, by the Brazilian agencies CNPq and CAPES, the PRONEX initiative of Brazil's MCT, and a FAPERJ BBP grant.

Valmir C. Barbosa

Berkeley, California
April 2000

is generally devoted to a discussion of the main properties and applications of SER.

Chapter 1 contains introductory material, such as a discussion of the origins of SER in the field of computer networks and a first glimpse into infinite sequences of acyclic orientations and some of those sequences' properties. This chapter also introduces the notion of neighborhood constraints, which establishes the basis for the treatment in Chapter 2 of the main application areas of SER. These can be broadly categorized into applications to resource-sharing computations and to computations on certain networks of automata, both discussed in Chapter 2.

Chapters 3 and 4 contain a discussion of the known properties of the SER dynamics. Chapter 3 is devoted to the properties that are well established, including properties related to concurrency. Chapter 4, by contrast, contains a collection of interesting observations, statistics, and correlations that may in the future inspire the creation of successful heuristics to optimize concurrency.

In addition to the publications that have contributed to the understanding of SER and of its properties, several other accounts of SER can now be found in the literature at various levels of detail. These include books [4, 5, 12, 20, 55], survey articles (e.g., [13]), and doctoral theses [3, 31]. The literature also contains accounts of how SER can be regarded as being built out of simple "synchronizing operations" [26, 57], in the sense of synchronizers for distributed algorithms [2], of SER in the context of the so-called "labeling problems" [1], and of SER-related systems in a stochastic setting [16].

1

Introduction

SER originated in the field of computer networks as a mechanism to rebuild communications routes after topological changes [32], as we discuss in Section 1.1. In this context, a finite number of sink-to-source reversals takes place (until routes are rebuilt). Most SER applications, however, require sequences of reversals that can be extended for as long as needed, so that agents can be scheduled for action "infinitely often." In Section 1.2, we discuss infinite sequences of acyclic orientations, of which the sequence that results from SER is a special case.

1.1. SER origins and neighborhood constraints

Suppose that G represents a computer network in such a way that each node stands for a message router and each edge for a bidirectional communications channel. Suppose further that G is oriented acyclically by an orientation ω such that one single node, say n_0, is a sink. It follows from the existence of one single sink in ω that every node lies on some directed path to that sink, and consequently ω can be regarded as providing all routers in the network with at least one path to the router represented by n_0.

Should the network undergo a topological failure (that is, the loss of a communications channel or of a router, which can be represented by the elimination of all communications channels attached to that router), it may still be possible for all routers to send messages to n_0, so long as at least one directed path remains from each router to n_0. However, it may happen that

Although we pursue the topic no further in this book, the reader should bear in mind that this simple implementation of the basic edge-reversal mechanism is resilient to any degree of "asynchronism," in the sense normally employed in the field of distributed algorithms [5]. What this means is that the time bases for the several agents can be totally independent of one another, and that it may take a message any finite amount of time to be delivered between two agents that are directly connected to each other in G.

Naturally, this simple implementation is dependent upon the existence of appropriate initial conditions, that is, of an acyclic orientation already established on G. Establishing such an orientation is subject to the optimization criteria to be introduced in Chapter 3, which of course poses implementational difficulties of its own. If, in practice, nodes can be assumed to have totally ordered identifications, then the establishment of an acyclic orientation without any concern to the aforementioned optimization criteria is a trivial matter: one may, for example, let all edges be oriented from the node with the highest identification to the one with the lowest. If such an assumption cannot be made, that is, if the system is of the so-called "anonymous" type [5], then one must resort to randomized procedures for the establishment of the initial acyclic orientation [17, 28, 29].

Another crucial element of SER-related computations that immediately becomes apparent from our discussion so far is that no two nodes ever become sources from sinks at the same time step if they are connected to each other by an edge in G. This comes from the obvious fact that no two sinks can be connected by an edge, and forms the basis for all the applications of SER that we address in Chapter 2. In all those applications, G is built out of a *neighborhood* on the node set N, that is, an irreflexive, symmetric binary relation on N, and nodes that are related to each other by this relation are called *neighbors*. This neighborhood relation gives rise to the set E of G's edges, and its importance, in all of those applications, is that agents are precluded from acting at the same time step if they are neighbors.

We use the term *neighborhood constraint* to refer to the prohibition that neighboring agents act simultaneously. Likewise, the term *neighborhood-constrained system* refers to each of the applications requiring that constraint. Neighborhood-constrained systems constitute another motivation for our overall assumption that G is connected. For graphs with multiple connected components, neighborhood constraints only make sense within each connected component.

1.2. Sequences of acyclic orientations

The sequences of acyclic orientations that result from the route-recovery mechanism of Section 1.1 are always finite: they end when node n_0 is the only sink, because n_0 is never turned into a source. Clearly, if n_0 were allowed to be turned into a source, then the turning of at least one sink into a source to obtain the next orientation would yield an infinite sequence of acyclic orientations.

From the standpoint of the neighborhood-constrained systems that we treat in Chapter 2, such infinite sequences are the ones that matter. At each time step $s \geq 0$, the sinks of ω_s that are sources in ω_{s+1} are the agents allowed to act at time s. These sinks constitute an *independent set* of G, that is, a subset of N that contains no neighbors.

An infinite sequence of acyclic orientations such that, for $s \geq 0$, ω_{s+1} is obtained from ω_s by turning at least one of the sinks in ω_s into a source is called a *schedule*. It is called a *greedy schedule* if the sinks that are turned into sources to yield ω_{s+1} are all the sinks of ω_s. Note that, as defined, SER refers to greedy schedules only, although it shares with all other schedules the basic mechanism of turning sinks into sources.

If, for two orientations ω and ω', a schedule exists containing both orientations, and such that ω precedes ω' in the schedule, then we say that ω' is *reachable* from ω. Characterizing the reachability of acyclic orientations of G from one another can be done as follows. For any two acyclic orientations ω and ω', and for $n_i \in N$, let $m_i(\omega, \omega')$ denote the number of times that n_i must be turned from sink into source in order for ω' to be obtained in some schedule from ω.

Now, for all edges (n_i, n_j) such that $\omega\big((n_i, n_j)\big) = n_j$, consider the equations

$$m_j(\omega, \omega') = \begin{cases} m_i(\omega, \omega'), & \text{if } \omega\big((n_i, n_j)\big) = \omega'\big((n_i, n_j)\big); \\ m_i(\omega, \omega') + 1, & \text{otherwise.} \end{cases} \qquad (1.1)$$

What these equations are saying is that, in order for ω' to be reachable from ω, neighbors n_i and n_j must be turned from sinks into sources an equal number of times (if the edge (n_i, n_j) is oriented likewise by both ω and ω'), or n_j must be turned from sink into source once more than n_i (otherwise, and assuming that ω orients (n_i, n_j) in the direction of n_j).

Solving the system of e equations (1.1) requires n nonnegative integers to be found, each representing the number of times the corresponding node is to be turned from a sink into a source in order to obtain ω' from ω. As it turns out [3], this can be done if and only if ω' is reachable from ω. In particular, every orientation is reachable from itself, and the corresponding solutions to (1.1) are all (and only) n-tuples having the same nonnegative integer in all entries.

For G the 5-node ring, we provide an illustration in Figure 1.2. This figure shows ten acyclic orientations, together with all the possible transitions in and out of them. Transitions depicted by solid lines are those that occur by turning all sinks in the orientation of origin into sources; dashed lines depict the remaining transitions. Greedy schedules in this figure are all the infinite sequences of orientations that can be traced on solid lines only. These are the SER schedules, which clearly do not suffice to realize the reachability of every orientation from itself.

We finalize the chapter by stating two additional properties of schedules [8]. Let ω_0 be any acyclic orientation, and consider the greedy schedule that starts at ω_0. For this schedule, and for any time step $s > 0$, let $m_i(s)$ denote the number of times node n_i is a sink in the sequence $\omega_0, \ldots, \omega_{s-1}$. Now consider any schedule starting also at ω_0 (possibly the greedy schedule, but not necessarily), and let $m_i'(s)$ denote the number of times n_i is turned from sink into source in the sequence $\omega_0, \ldots, \omega_s$. For the greedy schedule, $m_i'(s) = m_i(s)$ for all $s > 0$. For the general case, we have

$$m_i'(s) \leq m_i(s)$$

for all $n_i \in N$ and all $s > 0$, meaning that the greedy schedule from a given orientation allows nodes to turn from sources into sinks at least as frequently as any other schedule.

The second property is valid for all schedules and has important consequences for our analysis of SER in Chapter 3. If, for nodes n_i and n_j, we let r_{ij} denote the distance between n_i and n_j in G (this is the number of undirected edges on the shortest path between the two nodes), then we have, for all $s > 0$,

$$\left| m_i'(s) - m_j'(s) \right| \leq r_{ij}. \tag{1.2}$$

What this means is that the basic mechanism of turning sinks into sources precludes nodes from doing it too far ahead of one another. This, incidentally, is what guarantees, in the context of Section 1.1, that the route-recovery process stops when n_0 becomes the only sink in the graph: all further activity by the other nodes is dependent upon the turning of n_0 into a source, which never happens.

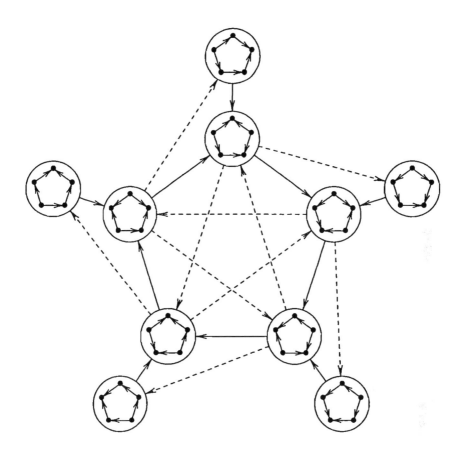

Figure 1.2. Some schedules for the 5-node ring

2

Neighborhood-Constrained Systems

In this chapter, the set N of G's nodes is regarded generically as a set of agents. An agent must perform certain actions from time to time, but must never do so concurrently with any of its neighbors in G, that is, other agents to which it is connected by an edge in E. Graph G is then representative of a neighborhood-constrained system, as introduced in Section 1.1. We assume, without any loss in generality, that the system that G stands for is driven by a time basis that is common to all agents. Also, we assume that performing an action takes no time and that messages sent between agents that are neighbors take no longer than the duration of one time unit to be delivered. Relaxing these assumptions to bring the system into a more realistic setting is a simple matter [5].

For systems with these characteristics, SER can be used to schedule agents to perform actions rather simply, as follows. For $s \geq 0$, the agents that are allowed to perform actions at time s are those that are sinks in ω_s, all nonneighbors therefore. In addition to providing the basic mechanism to respect the system's neighborhood constraints, SER also guarantees that agents can perform their actions approximately equally frequently whenever needed. This is a requirement for all the systems reviewed in this chapter, and is guaranteed, in essence, by (1.2).

We discuss neighborhood-constrained systems within two broad categories. First, in Section 2.1, we discuss systems for resource sharing by computational processes. Then we turn in Section 2.2 to networks of automata with applications to problems in artificial intelligence and in optimization.

2.1. Resource-sharing systems

One of the areas to which SER was first found to be applicable as an agent scheduler was the area of resource sharing in computer systems when the requirement of *mutual exclusion* for the sharing of resources holds (that is, a resource must be used by at most one process at a time). In this setting, N is a set of processes and an edge in E connects two processes if and only if those processes may at some time share a resource. For the sake of simplicity, we consider computations in which a process requires access to all the resources it shares with other processes whenever it requires access to any resource. Because mutual exclusion must be enforced, neighbors never access shared resources concurrently in such systems.

This system is a generalization of the paradigmatic *dining philosophers problem* [24], which considers five philosophers sitting at a round table, each two neighbors sharing a fork between them. A philosopher alternates periods of thinking and eating. Eating periods start when the philosopher has acquired the two forks adjacent to him, having become hungry, and end after a finite period of time. The problem is to devise a fork-sharing policy that will allow philosophers to eat within a finite time of becoming hungry, given that every fork must be used exclusively by one philosopher at a time.

For the dining philosophers problem, G is the 5-node ring, each node representing a philosopher and each edge representing a fork. In the generalized form of the problem, G is any connected graph, whose nodes and edges can continue to be regarded as philosophers and forks, respectively. In this case, a philosopher stands for a process and a fork stands for all the shared resources that the corresponding edge represents. Because of the restricted type of computations we have assumed, acquiring a fork is equivalent to acquiring all the resources that the corresponding processes share.

In addition to ensuring mutual exclusion, an acceptable solution to this resource-sharing problem must also ensure the absence of *deadlocks* and of *lockouts* at all times. That is, the solution must be such that at all times at least one process requiring access to resources succeeds in accessing them (no deadlock), and all processes succeed in accessing resources within a finite time of the need for those resources having arisen (no lockout). Note, as a curiosity, that this no-lockout condition is also known as *starvation-freedom*, owing to the dining-philosopher abstraction often used to analyze such systems.

The use of SER as a basis to a solution to the generalized dining philosophers problem goes along the following lines [19]. A philosopher that becomes hungry sends requests for all forks that he lacks to the neighbors who have them. A request for a fork is granted immediately, unless the philosopher receiving the request is using the fork, in which case he grants access to the fork immediately upon finishing, or unless that philosopher is also hungry and the edge connecting the requesting philosopher to him is oriented toward himself. In the latter case, granting the request is postponed until after the philosopher has acquired all forks that he himself lacks and has eaten. After eating, a philosopher reverses the orientation of all edges oriented toward himself.

Note, first of all, that the orientation of the edges of G is in this solution used to establish a priority structure on the set of philosophers. When a conflict arises between two hungry philosophers for the same fork, priority is given to the philosopher toward which the corresponding edge is oriented. If the orientation is acyclic, then at all times at least one hungry philosopher must get to eat (a sink, in the worst case). Also, every philosopher gets to eat within a finite time of having become hungry, because the worst that can happen is a wait for all philosophers on directed paths from him to sinks to eat. This solution, therefore, has the potential to guarantee the desired properties of deadlock- and lockout-freedom.

But it remains to argue that the changes done to acyclic orientations guarantee acyclicity. This argument is necessary because those changes are not the turning of sinks into sources that SER employs, but rather the turning of any node into a source. Arguing for sustained acyclicity is simple, though: as in the case of SER, a resulting directed cycle would have to go through the node or nodes where the changes in orientation were effected, which is impossible because what those changes do is to turn those nodes into sources.

Although this solution does not employ SER directly, but rather a variation thereof, there exists a more restricted form of the generalized dining philosophers problem to which SER can be applied directly. In this restricted version, called the *heavy-load* case of the problem, all philosophers are perennially hungry, but do nevertheless refrain from eating continually out of the need to let the other philosophers eat as well. In this case, the solution we outlined is such that only philosophers that are sinks eat, and the evolution of acyclic orientations of G takes place exactly according to SER, that is, all sinks are turned into sources.

To finalize, we note that a variation of this heavy-load case exists that requires philosophers to eat at relative frequencies that are not "uniform" in the sense that SER guarantees. For that variation, a scheduling mechanism similar to SER has been proposed [6, 30], but we do not pursue it any further in this book.

2.2. Partially concurrent networks of automata

Another interesting class of neighborhood-constrained systems comes from letting N be a set of automata and E a neighborhood relation that depends intimately on the specific automata at hand and is for now left unspecified. Each automaton n_i is responsible for updating an integer variable v_i periodically, taking values from a finite domain $D = \{0, \ldots, d-1\}$. This updating is done as a function of the variable's current value and of the values of the variables of the automaton's neighbors. This *updating function* is denoted by f_i, having the form $f_i : D^{1+x_i} \to D$, where x_i is the number of neighbors of n_i. The updating of v_i by n_i takes place via the assignment

$$v_i := f_i\big(v_i, v_j; \ (n_i, n_j) \in E\big). \tag{2.1}$$

The function f_i is the same for all $n_i \in N$, differing only in the number of arguments it takes, but not in how it depends on them. For this reason, we henceforth denote it simply by f.

In (2.1), the notation $v_j; \ (n_i, n_j) \in E$ stands for "all v_j such that n_i and n_j are neighbors." Similarly, throughout the section the notation $v_j = d_j; \ Q$ for some $d_j \in D$ and some predicate Q is to be understood as "$v_j = d_j$ for all v_j such that Q."

Graph G can be regarded as a network of automata, and depending on the specifics of each particular network, the updating function f can be of a deterministic or probabilistic nature. In all cases to be considered in this section, and for reasons that will be discussed when addressing each specific case, no two variables may be updated simultaneously if they belong to neighbors, and all variables must be updated *infinitely often*, in the following sense. There has to exist an integer $S > 0$ such that, for all $s \geq 0$, all variables get updated within the time interval $s, \ldots, s + S$. Clearly, SER can be used to schedule nodes to update their variables by letting sinks be the nodes that may update their variables at each time. That the desired properties are guaranteed comes from the fact that sinks are never neighbors and from (1.2). These automaton networks are called *partially concurrent*, in allusion to the fact that parallelism in variable updating is at all times limited [4].

The first network that we discuss is the *binary Hopfield neural network* [42]. In this network, each node is viewed as an artificial neuron of binary state v_i, so $D = \{0, 1\}$. Associated with every two neurons n_i, n_j are the *synaptic strengths* w_{ij} and w_{ji}, related respectively to the influence of n_j upon n_i and of n_i upon n_j. These two neurons are neighbors if and only if either $w_{ij} \neq 0$ or $w_{ji} \neq 0$.

The updating function for this network is

$$v_i := \begin{cases} 0, & \text{if } \sum_{j=0}^{n-1} w_{ij}v_j + e_i \leq \theta_i; \\ 1, & \text{otherwise,} \end{cases} \qquad (2.2)$$

where e_i represents a fixed external input to the neuron and θ_i is the neuron's *threshold*. What is remarkable about (2.2) is that, if $w_{ij} = w_{ji}$ and $w_{ii} = 0$ for all $n_i, n_j \in N$, and if all neurons get to update their states infinitely often, then gradient descent is performed on the function

$$H = -\frac{1}{2} \sum_{i=0}^{n-1} \sum_{j=0}^{n-1} w_{ij}v_iv_j - \sum_{i=0}^{n-1} e_iv_i + \sum_{i=0}^{n-1} \theta_iv_i, \qquad (2.3)$$

if neighbors are not allowed to update their states simultaneously [42]. A simple example can be used to show that this prohibition on having neighbors update their states simultaneously is also necessary [4], which characterizes the binary Hopfield network as indeed a neighborhood-constrained system [9].

The quantity in (2.3) is often referred to as the network's *energy*. Performing gradient descent on this energy is important because it hints at several possible applications of the network. One example is as a means to implement content-addressable memories by associating certain local minima of H with items in $\{0,1\}^n$ that were stored in the network and need to be retrieved [40].

Another important area of application that the gradient-descent capability of these networks hints at is the area of combinatorial optimization, because neurons' states can be regarded as discrete variables on which an objective function having the form of (2.3) has to be minimized. Of course, the convergence to local minima that gradient descent elicits is no longer the boon that it was in the implementation of content-addressable memories, but is in general a hindrance instead, because what one seeks are global minima.

Fortunately, it is possible to turn the updating function in (2.2) into a time-varying probabilistic rule that allows for global minima of H to be reached under certain assumptions. In order to do that, we first rewrite (2.2) as

$$v_i := \begin{cases} 0, & \text{if } -\Delta_i H \leq 0; \\ 1, & \text{otherwise,} \end{cases} \qquad (2.4)$$

where $\Delta_i H$ denotes the difference from the value of H when $v_i = 0$ to the value of H when $v_i = 1$, all other variables kept constant. Viewed as in (2.4), the updating function can then be rewritten as the following probabilistic rule, as long as the parameter T in this rule approaches zero. Assign 0 to v_i with probability

$$\frac{1}{1 + \exp(-\Delta_i H/T)},\tag{2.5}$$

otherwise assign 1. If, on the other hand, T has a nonzero value, then the rule based on (2.5) is no longer equivalent to (2.4), but rather allows the gradient-descent behavior of the binary Hopfield network to be replaced with occasional "uphill" moves. The parameter T of (2.5) is the temperature parameter of simulated annealing [48], which, if lowered gradually from an initial high value, gives rise to a time-varying updating function based on (2.5) that may lead to convergence to global (as opposed to local) minima of H.

This probabilistic variation of the binary Hopfield network is known as the *Boltzmann machine* [41], and its characterization as a neighborhood-constrained system proceeds along the following more general lines. Let V be the set of variables $\{v_0, \ldots, v_{n-1}\}$, and consider a probability distribution P over D^n. Consider also a symmetric neighborhood $\mathcal{N}$ on the set V. Two definitions are in order [14, 37, 38, 44, 47, 60].

We say that V is a *Markov random field (MRF)* with respect to P and to $\mathcal{N}$ if and only if P is strictly positive everywhere and, in addition,

$$P(v_i = d_i \mid v_j = d_j;\ v_j \neq v_i) = P(v_i = d_i \mid v_j = d_j;\ (v_i, v_j) \in \mathcal{N}).\tag{2.6}$$

The positivity requirement on P can be done away with many times [52], but (2.6) is what matters: it states that the neighborhood of a variable is as good a condition as the entire remainder of the set V.

Likewise, the set V is said to be a *Gibbs random field (GRF)* with respect to P and to $\mathcal{N}$ if and only if P is the *Boltzmann-Gibbs distribution*, that is,

$$P(d_0, \ldots, d_{n-1}) = \frac{\exp\big(-H(d_0, \ldots, d_{n-1})/T\big)}{Z},\tag{2.7}$$

where T is a temperature parameter, Z is a normalizing constant, and H is any function that can be expressed as a sum of components, each of which cannot itself be broken into a sum and depends exclusively on the variables of a subset of V that is completely connected by $\mathcal{N}$. Clearly, the energy of the Hopfield network as given by (2.3) is a function of this type for suitable $\mathcal{N}$.

A very deep and remarkable connection exists between MRFs and GRFs: V is an MRF with respect to P and to $\mathcal{N}$ if and only if it is also a GRF with respect to P and to $\mathcal{N}$. While it may be relatively simple to see that every GRF has to be an MRF, the opposite direction is far from trivial to prove [4, 37, 38].

The computation of interest over these random fields is the so-called *Gibbs sampling*. This computation updates the value of each variable v_i according to probabilities conditioned upon all other variables' values, which by (2.6) is equivalent to conditioning upon the values of the variable's neighbors. By (2.7), v_i is updated to a value $d_i \in D$ with probability

$$\frac{1}{1 + \sum_{d_i' \neq d_i} \exp(-\Delta_i H/T)}, \qquad (2.8)$$

where $\Delta_i H$ is the difference from the value of H when $v_i = d_i$ to the value of H when $v_i = d_i'$, all other variables remaining constant. The probability in (2.8) generalizes the probability in (2.5).

If we now return to the graph G and let its set of edges E be given to match the neighborhood $\mathcal{N}$ in the obvious way, then we see that the process of Gibbs sampling is indeed another computation of the general type we have been considering in this section, that is, a computation in which the automata in N update their variables based on information from their neighbors. Specifically, $\Delta_i H$ depends only on n_i's variable and those of its neighbors, so updating functions based on (2.8) are indeed of the type given by (2.1).

There are two properties of Gibbs sampling that are of interest to us [34]:

- If Gibbs sampling is performed infinitely often with constant T and no neighbors are allowed to update their variables simultaneously, then the distribution one observes as time progresses converges to the distribution of (2.7).

- If Gibbs sampling is performed infinitely often with varying T (decreasing from a high initial value, but no faster than a certain rate that we leave unspecified) and no neighbors are allowed to update their variables simultaneously, then the distribution one observes as time progresses converges to the uniform distribution over those points in D^n at which the function H is globally minimum.

Once again, it takes only a simple example to demonstrate that forbidding simultaneous updates by neighbors is also necessary for the above two properties to hold [4]. So long as one wishes to have the guarantees offered by these properties, the process of Gibbs sampling is seen to give rise to a neighborhood-constrained system. Respecting the associated neighborhood constraints is how optimization by simulated annealing should be parallelized [7], although there are accounts of implementations that have ignored these constraints and yet appear to have yielded acceptable results [23, 27, 35, 36].

As one last example of networks of automata that can exhibit the neighborhood-constraint property, we turn to *Bayesian networks* [55]. A Bayesian network is an acyclic directed graph whose vertex set is the set of variables $V = \{v_0, \ldots, v_{n-1}\}$ and whose directed edges represent direct causal influences of one variable upon another. Associated with each variable v_i are the set $\mathcal{P}_i$ of its *parents* (immediate ancestors in the graph), the set $\mathcal{C}_i$ of its *children* (immediate descendants in the graph), and the set $\mathcal{M}_i$ of its *mates* (the other parents of its children). Also, assuming without any loss in generality that all variables are binary (i.e., they take values from the set $D = \{0, 1\}$), then associated with v_i are also the $2^{1+|\mathcal{P}_i|}$ probabilities that v_i takes on one of the two possible values conditioned upon the values of its parents, that is,

$$P(v_i = d_i \mid v_j = d_j; \ v_j \in \mathcal{P}_i). \tag{2.9}$$

Under the assumption that every variable is independent of its nondescendants in the directed graph given its parents [55], it can be shown very easily that the joint distribution over the set V is given by the product of the probabilities in (2.9), that is,

$$P(d_0, \ldots, d_{n-1}) = \prod_{i=0}^{n-1} P(v_i = d_i \mid v_j = d_j; \ v_j \in \mathcal{P}_i). \tag{2.10}$$

The most essential problem that one has to solve in connection with Bayesian networks is the problem of computing the *joint posterior distribution* given a set of evidences $\mathcal{E} \subseteq V$. Although this distribution can be derived analytically in a rather straightforward manner from (2.10) [53], assessing the probability of a single variable in $V \setminus \mathcal{E}$ given the variables in $\mathcal{E}$ is computationally intractable [22], in the sense of NP-completeness [33].

One of the approximation techniques that aims at overcoming this limitation is the technique of *stochastic simulation* [54]. What it does is to update all variables infinitely often based on conditional probabilities given the values of all other variables. Similarly to what happens with MRFs, it is not really necessary to consider the values of all other variables. Specifically, variable $v_i \in V \setminus \mathcal{E}$ is assigned value d_i with probability

$$\alpha P(v_i = d_i \mid v_j = d_j; \ v_j \in \mathcal{P}_i) \prod_{v_k \in \mathcal{C}_i} P(v_k = d_k \mid v_\ell = d_\ell; \ v_\ell \in \mathcal{P}_k), \tag{2.11}$$

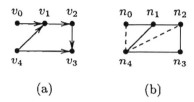

Figure 2.1. A Bayesian network (a) and the resulting graph G (b)

where α is a normalizing constant. By (2.11), it is apparent that the variables that matter are those in the set $\mathcal{L}_i = \mathcal{P}_i \cup \mathcal{C}_i \cup \mathcal{M}_i$.

As it turns out, the best way to analyze the properties of stochastic simulation is to recognize that the set $V \setminus \mathcal{E}$ is a GRF with respect to the neighborhood induced by the sets $\mathcal{L}_i$ for all $v_i \in V$ [43]. For this GRF, $T = 1$ and

$$H = -\sum_{i=0}^{n-1} \ln P(v_i = d_i \mid v_j = d_j;\ v_j \in \mathcal{P}_i), \qquad (2.12)$$

where the completely connected subsets of V upon which the components of H depend are $\{v_i\} \cup \mathcal{P}_i$ for all $v_i \in V$. The Boltzmann-Gibbs distribution, conditioned upon the variables in $\mathcal{E}$, that results from (2.12) is the joint posterior distribution derived in [53].

What this characterization as a GRF means is that a network of automata can be set up to do Gibbs sampling on the set $V \setminus \mathcal{E}$, employing for such the conditional probabilities in (2.11) for nodes to update their variables. By the properties of Gibbs sampling when it is performed infinitely often on such variables and respecting the neighborhood constraints, convergence to the joint posterior distribution is guaranteed. In this network of automata, the set E of undirected edges mirrors the GRF neighborhood. This is illustrated in Figure 2.1, where a Bayesian network and the resulting graph G are shown. Dashed lines in part (b) represent the edges that in G join nodes whose variables are not joined by a directed edge, in any of the two possible directions, in the Bayesian network. These variables are mates of one another.

In the same vein, if during the Gibbs sampling on this GRF we let T vary as prescribed earlier, then we know that a global minimum of (2.12) is eventually obtained. But this global minimum corresponds to a global maximum of (2.10), and also of the joint posterior distribution [53], so this varying-T version of Gibbs sampling can be used to identify a global maximum of the joint posterior distribution. For such, it suffices that each variable $v_i \in V \setminus \mathcal{E}$ be assigned value d_i with probability

$$\alpha_T P(v_i = d_i \mid v_j = d_j; \ v_j \in \mathcal{P}_i)^{1/T} \prod_{v_k \in \mathcal{C}_i} P(v_k = d_k \mid v_\ell = d_\ell; \ v_\ell \in \mathcal{P}_k)^{1/T},$$

$$(2.13)$$

where α_T stands for the normalizing constant given the value of T [43]. In conclusion, under stochastic simulation employing updating functions based on either (2.11) or (2.13), a Bayesian network is a neighborhood-constrained system [25, 55].

3

Scheduling by Edge Reversal

We return, in this chapter, to the analysis of SER that we started in Section 1.2. The chief concept to be discussed will be that of concurrency as it relates to SER, that is, the notion of how much parallelism can be achieved under SER. This notion is of crucial importance because it underlies the usefulness of SER as a practical method for scheduling the agents of a neighborhood-constrained system, and also because it relates closely to several of the chromatic indicators of G, thereby providing a strongly principled approach to the analysis of the SER dynamics.

This chapter's material is divided into two sections. Section 3.1 contains a discussion of the main properties of SER, in direct continuation of the material introduced in Section 1.2. Special cases of G for which further properties are known are also discussed. The notion of concurrency is formally introduced in Section 3.2, where its properties are analyzed from the perspective of both its combinatorial nature and its relation to the dynamics of SER.

Unless otherwise noted, this chapter is based on [3, 8].

3.1. Main properties and special cases

We start by recalling from Section 1.2 that a greedy schedule is any infinite sequence of acyclic orientations $\omega_0, \omega_1, \ldots$ such that, for all $s \geq 0$, ω_{s+1} is obtained from ω_s by turning all sinks into sources. Because we only discuss greedy schedules henceforth in the book, in the sequel we refer to them simply as schedules.

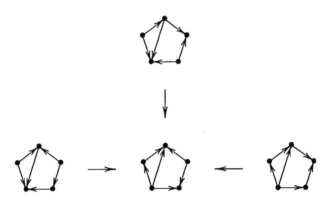

Figure 3.1. Possible precursors of an orientation

In any schedule, and for any $s \geq 0$, orientations ω_s and ω_{s+1} are such that a sink in ω_{s+1} has at least one neighbor that is a sink in ω_s. Now consider the question of determining the possible immediate precursors ω_s of ω_{s+1} in all schedules in which ω_{s+1} participates. In other words, the question is to determine all the orientations from which ω_{s+1} can be obtained by turning all of each orientation's sinks into sources. By the property we just described, every sink in ω_{s+1} has at least one neighbor that is a source also in ω_{s+1}. So the question can be answered by considering all the nonempty subsets of G's sources according to ω_{s+1} that dominate (are adjacent to) all sinks. Then ω_{s+1} has as many possible immediate precursors as there are subsets like this. If no such subset exists for a given orientation, then it can only exist in a schedule as the schedule's first orientation.

This is depicted in Figure 3.1, where the three possible immediate precursors of the center orientation are shown. Of these, the rightmost orientation has itself no precursors, because its single source and sink are not neighbors.

Because the SER rule to move from an orientation to the next is deterministic, and considering that the set Ω of all the acyclic orientations of G is finite, every schedule must eventually become periodic, that is, reach a group of orientations that will be indefinitely re-visited as time progresses. We call this group of orientations a *period*, and denote by $p(\omega)$ the number of orientations in the period that is reachable from orientation ω. We call $p(\omega)$ the period's *length*.

In addition to $p(\omega)$, another important characteristic of a period is the number of times each node is a sink in orientations of the period, called *periodic* orientations. It turns out that this number is the same for all nodes, as a direct consequence of (1.2): if any two nodes were sinks different numbers of times in a period, then all schedules going through that period would eventually violate that inequality. For the period that is reachable from orientation ω, we denote this number by $m(\omega)$. The two integers $m(\omega)$ and $p(\omega)$ have important roles to play in the remainder of the book. Whenever safe from the occurrence of ambiguities, we use m and p instead, respectively.

This periodic behavior of SER schedules was already apparent in Figure 1.2, which should be re-examined now. What that figure shows is a period of length $p = 5$ and $m = 2$ for all the orientations in the figure, periodic or otherwise. Note also in this figure that the orientations that it depicts are all the orientations from which that period can be reached. All other orientations of the 5-node ring, therefore, must be such that different periods are reachable from them.

What SER does then is to partition the set Ω into several *basins of attraction*, each comprising a period (the *attractor* or *limit cycle*) and all the other orientations from which that period can be reached. Each basin is characterized by its own m and p values, being part of a rich graph dynamics, in the sense of [56], that arises from the very simple rule of turning all sinks into sources.

Although several of an SER basin's properties are known and well characterized, there still remain open questions. One of them is the question of deciding, without recourse to the SER dynamics, whether a given acyclic orientation is periodic or not. Of the known properties, two interesting ones are the following:

- Let an acyclic orientation be called *transient* if it is not periodic. Then the number of transient orientations in any schedule is $O(n^4)$ [49].

- By contrast with this polynomial bound on the number of transient orientations of a schedule, there exist periods whose lengths are exponential in a quantity that is asymptotically bounded from below by $\sqrt{n}$ [50].

Next we describe another of the interesting properties of SER basins, but first additional concepts and notation are needed. For an acyclic orientation ω, the *sink decomposition of G according to ω* [11] is a partition of N into the $\lambda(\omega)$ independent sets $S_0, \ldots, S_{\lambda(\omega)-1}$ having the properties that S_0 is the set of sinks according to ω, and that, for $0 < \ell \le \lambda(\omega) - 1$,

S_ℓ is the set of sinks of the graph that remains after the nodes in all of $S_0, \ldots, S_{\ell-1}$ have been removed from G. We call $\lambda(\omega)$ the *length* of the sink decomposition of G according to ω, denoting it simply by λ when no ambiguities are possible.

A graph's sink decomposition has interesting properties. For example, every edge is oriented from a higher- to a lower-subscript set in $S_0, \ldots, S_\lambda$. Similarly, every node in S_ℓ has at least one neighbor in $S_{\ell-1}$, for $0 < \ell \leq \lambda - 1$. It follows from these two properties that $\lambda(\omega) - 1$ is the length of the longest directed path in G when it is oriented by ω. The following is another property of SER basins:

- Let ω and ω' be acyclic orientations of G such that ω' is obtained from ω by turning all sinks into sources. Then $\lambda(\omega) - 1 \leq \lambda(\omega') \leq \lambda(\omega)$.

What this property is stating is that the length of a graph's sink decomposition is nonincreasing along all possible schedules. In addition, whenever the length decreases, it does so by exactly one unit. One consequence of this is that all periodic orientations in the same basin must have sink decompositions of the same length.

We provide an illustration of this property in Figure 3.2, from which the reason why it is true becomes obvious, as follows. SER can be regarded as "shifting" the sets $S_1, \ldots, S_{\lambda-1}$ one step in the direction of lower subscripts, yielding the new sink decomposition $S'_0, \ldots, S'_{\lambda'-1}$ in such a way that $S'_0 = S_1$, $S'_1 \supseteq S_2$, and so on, through $S'_{\lambda-2} \supseteq S_{\lambda-1}$. When this happens, it may be that no set $S'_{\lambda-1}$ is formed, in which case $\lambda' = \lambda - 1$, or it may happen that $S'_{\lambda-1}$ is indeed formed, in which case it comprises nodes from S_0 only and $\lambda' = \lambda$. These are the situations shown, respectively, in parts (a) and (b) of the figure.

Sink decompositions are also useful in characterizing some periodic orientations, despite the fact that such a characterization is, as remarked earlier, generally unknown. The periodic orientations that we know how to characterize are those for which $m = 1$. Specifically, an acyclic orientation ω is periodic with $m(\omega) = 1$ if and only if every node in S_0 has at least one neighbor in $S_{\lambda(\omega)-1}$. In this case, $p(\omega) = \lambda(\omega)$ and the sets of sinks along the period are exactly the independent sets of the sink decomposition. This is what happens, for example, in part (b) of Figure 3.2.

One obvious $m = 1$ case is the case of complete graphs, since in such graphs every node is connected to every other node. If G is a complete graph, then it has as many acyclic orientations as there are permutations of its nodes, that is, $n!$ orientations. The sink decomposition of G according to each of these orientations has length n and is periodic with $m = 1$ and $p =$

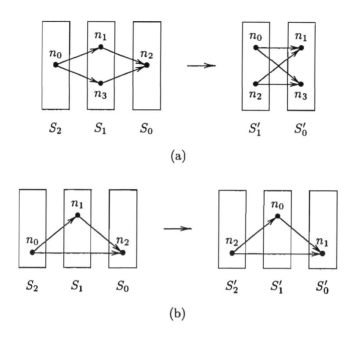

(a)

(b)

Figure 3.2. Evolving sink decompositions

n. Therefore, the set Ω of all acyclic orientations contains $(n-1)!$ basins, each comprising n periodic orientations and no transient orientations at all.

Figure 3.3 depicts all the basins of attraction for the 4-node complete graph. For the sake of visual clarity, no edges are shown in the figure, and the orientation of each edge is left to be inferred from the order in which the nodes are presented. If, for some orientation, n_i appears to the left of n_j for $0 \le i, j \le 3$, then edge (n_i, n_j) is oriented from n_i to n_j.

Another interesting example for which $m = 1$ for all basins is the case of trees. Unlike complete graphs, however, if G is a tree, then all of its 2^{n-1} acyclic orientations belong to the same basin of attraction in Ω. The corresponding period has length $p = 2$, and the independent sets in the sink decompositions of the two periodic orientations reflect the nature of trees as bipartite graphs, that is, graphs whose node sets can be partitioned into two independent sets. In other words, one of these independent sets is the set of sinks in one periodic orientation, the other in the other periodic

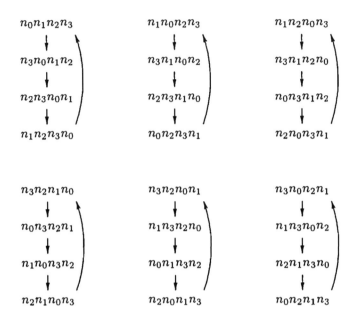

Figure 3.3. Basins of attraction for a complete graph

orientation. This is illustrated by Figure 3.4 for $n = 5$. Note that such $m = 1$, $p = 2$ basins happen not only for trees, but for all bipartite graphs, albeit in conjunction with other basins for nontrees.

Although the $m = 1$ case totally dominates Ω for complete graphs and trees, its occurrence is by no means restricted to those graphs. In fact, it is relatively easy to see that for every G there exists at least one basin for which $m = 1$. To see this, consider the following orientation for $e \geq n$ (the case of $e = n - 1$ is the case of trees, already analyzed). First locate the largest undirected simple cycle in the graph (a simple cycle is one that never goes through the same node more than once). On this cycle, choose any node to be a sink and any of its neighbors on the cycle to be a source, then orient all other edges on the cycle in the general direction from the source to the sink. Now work breadth-first from the single sink, orienting all untraversed and still not oriented edges inward to the node that is being visited. This will preserve the source chosen initially. It will also ensure that no source is created that is on some undirected cycle but

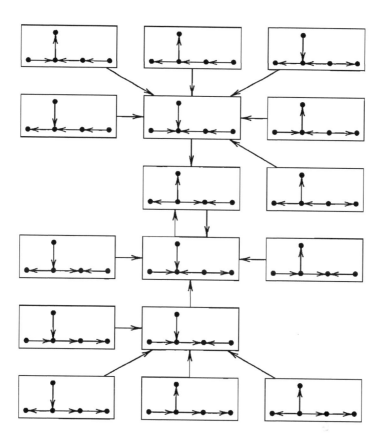

Figure 3.4. Basin of attraction for a tree

farther to the sink than that initial source. If every node were on some undirected cycle, then we would have one single sink adjacent to the far-thest source and be done. There may, however, be "tree appendages" to the graph, i.e., subgraphs whose edges are in no undirected cycle. As we work "outward" from the initial sink, sources may appear in these sub-graphs that are farthest to that sink than the initial source. This is easy to prevent, however, by simply creating new sinks whenever this would hap-pen. These sinks, like the initial one, will be adjacent to a source that is as far from the initial sink as the initial source. By our preceding discussion, the orientation whose construction we just outlined is periodic with $m = 1$.

We end this section with a discussion of rings. If G is a ring, then it has $2^n - 2$ acyclic orientations, and it is possible to characterize precisely which of those orientations are periodic. In order to do so, we look closely at certain patterns in the acyclically oriented ring, as follows. Given an acyclic orientation ω of the ring, let a *segment* be any maximal subset of contiguous nodes having even size and such that, according to ω, its nodes are alternately sinks and sources. If the ring has a nonempty segment according to ω, then the size of that segment is at least 2 and at most $2\lfloor n/2 \rfloor$. A segment of size $k > 0$ contains $k - 1$ edges that are internal to the segment, that is, edges that connect one node of the segment to another. The edge that connects the two extreme nodes of a size-n segment for n even is not considered internal, so, clearly, there always is at least one edge that is not internal to any segment.

We say that ω is *uniform* if and only if all edges that are not internal to any segment are oriented in the same overall direction (clockwise or counterclockwise). Because ω is acyclic, if it is uniform then the ring has at least one nonempty segment according to it. What is interesting about a uniform orientation, say one according to which noninternal edges are oriented clockwise, is that SER can be viewed as shifting the segments counterclockwise one edge per time step. What happens eventually is that this shifting pattern clicks back in place at its original position, meaning that the original uniform orientation is periodic. As it turns out, uniformity is also necessary for an acyclic orientation of a ring to be periodic. The length p of the period and the value of m depend on how the segments are arranged along the ring.

Some examples are given in Figure 3.5, where segments are enclosed in rectangular boxes. Periodic orientations are in part (a) of the figure, both having $m = 3$ and, from left to right, $p = 9$ and $p = 7$. Those of part (b) are transient.

This characterization of periodicity for rings has good intuitive appeal, but another, simpler one can easily be deduced from it. If the clockwise direction is the direction in which at least half the ring's edges are oriented by ω, then ω is periodic if and only if, when the ring is traversed in the clockwise direction, no two consecutive edges are encountered that are oriented counterclockwise.

3.2. Attractor dynamics and concurrency

As we have discussed so far in this chapter, the SER state space (the set Ω of all acyclic orientations of G) is partitioned into basins of attraction,

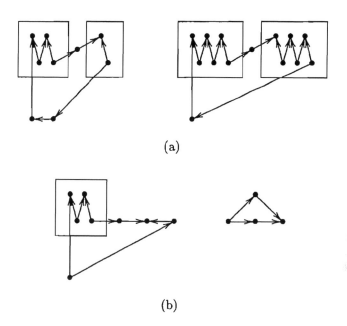

(a)

(b)

Figure 3.5. Periodic (a) and transient (b) orientations of rings

each one having a core period of orientations characterized by the integers m (how many times a node is a sink in periodic orientations) and p (how many periodic orientations there are). These two numbers can be thought of as characterizing the entire basin, because every schedule starting at some orientation in that basin is eventually attracted to the period lying at the basin's core.

From the point of view of using SER to schedule agents in a real application, such as those overviewed in Chapter 2, the crucial question is how to select the initial acyclic orientation so that the resulting schedule provides a good amount of parallelism or concurrency. Because of the attractor dynamics that governs SER, this amounts to a choice among basins of attraction, because the schedules, being infinite, are intuitively expected to have concurrency properties that depend chiefly on the basin's period rather than on its transient orientations. Furthermore, although a measure for how much concurrency is attained is still undefined, we should expect more concurrency for higher m and lower p, because this means that more nodes are sinks in a shorter interval of time.

Our measure for the concurrency attained along a schedule tries to capture precisely this expectation. It defines the amount of concurrency attained by the SER schedule from an initial orientation ω_0 as the average, over time and over n, of the number of times nodes have been sinks in the limit as time tends to infinity. This average is denoted by $\gamma(\omega_0)$ and is defined as

$$\gamma(\omega_0) = \lim_{s \to \infty} \frac{1}{ns} \sum_{i=0}^{n-1} m_i(s), \tag{3.1}$$

where $m_i(s)$, as we recall from Section 1.2, is the number of times node n_i is a sink in the first s orientations of the schedule. We use γ in place of $\gamma(\omega_0)$ whenever ω_0 is understood from the context. Not surprisingly, it follows from (3.1) that

$$\gamma(\omega_0) = \frac{m(\omega_0)}{p(\omega_0)}. \tag{3.2}$$

What (3.2) tells us is that the amount of concurrency that is attained by SER from ω_0 according to the definition in (3.1) depends exclusively on the period that is eventually reached from ω_0. It is therefore the same for all orientations in the same basin of attraction, and leads naturally to the question of whether such an amount of concurrency can be computed by examining the initial orientation alone, thereby bypassing any recourse to the SER dynamics apparently implied by (3.2). Of course, this question is moot if G is a complete graph or a tree, because in these cases we already know that $\gamma(\omega_0) = 1/n$ and $\gamma(\omega_0) = 1/2$, respectively, directly from (3.2) and from the knowledge of the graph's structure.

Let us then assume that G is not a tree (we still leave complete graphs in because, as will become clear shortly, what matters are the graph's undirected cycles). We proceed to answering the question affirmatively as follows. Let K denote the set of all simple cycles of G. For $\kappa \in$ K, let $e^+(\kappa, \omega_0)$ denote the number of edges in κ that are oriented in one of the two possible directions by ω_0. Likewise for $e^-(\kappa, \omega_0)$ and the other direction. Note that these two quantities are invariant under the turning of sinks into sources, because this operation adds to each of them as much as it subtracts from them. They are, therefore, the same along any schedule. Not only this, but they are also the same for all the orientations in the same basin of attraction. The reason for this is that all such orientations that are not part of a certain schedule in that basin can be obtained from orientations in the schedule by simply turning selected sources into sinks, which again adds as much as it takes.

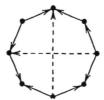

Figure 3.6. Assessing concurrency from simple cycles

Given these definitions, it can be shown that

$$\gamma(\omega_0) = \min_{\kappa \in K} \frac{1}{|\kappa|} \min\{e^+(\kappa, \omega_0), e^-(\kappa, \omega_0)\}, \qquad (3.3)$$

where $|\kappa|$ is the number of nodes in κ. The quantity in (3.3) can be computed in polynomial time [8, 49], which would not have been possible in general had we resorted to the dynamics of SER in search of the values of $m(\omega_0)$ and $p(\omega_0)$—as we recall from Section 3.1, there do exist exponentially large periods.

The alternative expression for concurrency given in (3.3) implies that, if G is not a tree, then the amount of concurrency that is achieved from a certain acyclic orientation depends on how that orientation affects the edges of the simple cycle (or cycles) for which

$$\frac{1}{|\kappa|} \min\{e^+(\kappa, \omega_0), e^-(\kappa, \omega_0)\}$$

is minimum over K. If κ^* is one of these critical simple cycles, and if the "+" direction is the one along which fewest of the cycle's edges are oriented, then the following also holds. Every contiguous portion of κ^* oriented in the "+" direction is a shortest directed path in G oriented by ω_0; likewise, every contiguous portion of κ^* oriented in the "−" direction is a longest directed path in G oriented by ω_0.

Consider, for the sake of illustration, the graph and acyclic orientation depicted in Figure 3.6. In this case, (3.3) yields $\gamma = 3/8$ along the simple cycle shown in solid lines when it is traversed clockwise.

Now let $\gamma^*(G)$ be such that

$$\gamma^*(G) = \max_{\omega \in \Omega} \gamma(\omega). \qquad (3.4)$$

By (3.3), $\gamma^*(G)$ is such that

$$\frac{1}{n} \leq \gamma^*(G) \leq \frac{1}{2}, \tag{3.5}$$

which is in accordance with the intuition regarding the functioning of SER we may have built from (3.2): it takes at most n time steps for a node to become a sink (so $m \geq p/n$), and the most frequently that a node becomes a sink is in every other time step (so $m \leq p/2$).

The problem of determining $\gamma^*(G)$ is the problem of finding the initial acyclic orientation of G that optimizes concurrency. In the remainder of this chapter, we discuss several issues related to this problem. First, however, a few definitions related to coloring the nodes of G are in order.

A *coloring* of the nodes of G is an assignment of natural numbers (*colors*) to nodes in such a way that no two neighbors are ever assigned the same number. The minimum number of colors needed to color the nodes of G with one color per node is the graph's *chromatic number*, denoted by $\chi(G)$. The determination of $\chi(G)$ is in general an NP-hard problem [33, 46]. If, for $k \geq 1$, the number of distinct colors to be assigned to each node is k, then the minimum number of colors needed overall to color the nodes of G is the *k-chromatic number* of G, denoted by $\chi^k(G)$. Colorings of G's nodes that allow more than one color to be assigned to each node are also referred to as *multicolorings* [21, 58, 62].

There is a sense in which a multicoloring can be said to be more or less "efficient" than another: a multicoloring that employs a certain number of colors overall per color assigned to each node is more "efficient" than another for which this ratio is larger. The minimum ratio is called the *multichromatic number* (or *fractional chromatic number*) of G, denoted by $\chi^*(G)$ and formally given by

$$\chi^*(G) = \min_{k \geq 1} \frac{\chi^k(G)}{k}. \tag{3.6}$$

The problem of determining $\chi^*(G)$ is in general NP-hard as well [39].

Node multicolorings are important in our present context because every SER period can be regarded as providing the nodes of G with a node multicoloring. If m and p are the integers characterizing the period, then this multicoloring assigns m colors to each node, employing for such p colors overall. The exact colors assigned to each node depend on the orientations within the period in which the node is a sink. If, for $t \geq 0$, $\omega_t, \ldots, \omega_{t+p-1}$ are the orientations in the period, then node n_i is assigned color u, for $0 \leq u \leq p - 1$, if and only if n_i is a sink in ω_{t+u}.

One interesting question is then whether other multicolorings exist that are more "efficient" than the SER multicoloring that yields $m/p = \gamma^*(G)$. Such a multicoloring would assign k colors to each node, employing q colors overall, in such a way that $q/k < p/m$, or, equivalently, $k/q > m/p$. A scheduling mechanism based on this other multicoloring would schedule independent sets sharing the same color cyclically, just as inside an SER period, and would provide more concurrency than SER, as $k/q > \gamma^*(G)$.

This question can be answered affirmatively, for example for the graph in Figure 3.6. For this graph, it can be shown that $\chi^*(G) = 5/2$ [62], which means that it is possible to schedule independent sets of nodes with more concurrency than SER can provide, given by $\gamma^*(G) = 3/8 < 2/5$ (one periodic orientation for which this SER concurrency is achieved is precisely the one shown in the figure).

Note, however, that the multicolorings implied by SER periods are all *interleaved multicolorings*, in the following sense. If, for two neighbors n_i and n_j, the colors assigned to n_i are $k_i^1 < \cdots < k_i^m$, while those assigned to n_j are $k_j^1 < \cdots < k_j^m$, then either $k_i^1 < k_j^1 < \cdots < k_i^m < k_j^m$ or $k_j^1 < k_i^1 < \cdots < k_j^m < k_i^m$. This follows from the fact that, in any SER schedule, neighbors always appear as sinks in alternating orientations. As it turns out, every interleaved multicoloring employing a total of q colors to assign k colors to each node is such that $\gamma^*(G) \geq k/q$. Consequently, multicolorings for which

$$\gamma^*(G) < \frac{k}{q} \leq \frac{1}{\chi^*(G)}$$

must not be interleaved. If we rewrite (3.4) as

$$\frac{1}{\gamma^*(G)} = \min_{\omega \in \Omega} \frac{1}{\gamma(\omega)},$$

then by analogy with (3.6) we can call $1/\gamma^*(G)$ the *interleaved multichromatic number* (or *interleaved fractional chromatic number*) of G.

Every coloring that assigns one single color to each node is also an interleaved multicoloring, so it follows from the discussion we just had and from viewing SER periods as node multicolorings that

$$\frac{1}{\chi(G)} \leq \gamma^*(G) \leq \frac{1}{\chi^*(G)}. \tag{3.7}$$

Together, (3.5) and (3.7) provide the known bounds on $\gamma^*(G)$. Note, however, that $\chi(G) \leq n$, so the lower bound in (3.7) is tighter than the one in

(3.5). Similarly, every multicoloring assigning k colors to each node must be such that $\chi^k(G) \geq 2k$ (this is the number of colors required by any two neighbors). So $\chi^*(G) \geq 2$, and consequently the upper bound in (3.7) is also tighter than the one in (3.5).

Let us then return to the question of how to find $\gamma^*(G)$. Because the inequality in (3.7) is sometimes strict (once again the graph in Figure 3.6 is an example, having $\chi(G) = 3$), there might in principle be hope that such a task could be carried out efficiently. However, a simple polynomial-time reduction from the decision problem related to finding $\chi(G)$ leads to the conclusion that the determination of $\gamma^*(G)$ is also an NP-hard problem in general. Nevertheless, the value of $\gamma^*(G)$ is known for some special graphs. We have $\gamma^*(G) = 1/2$ for bipartite graphs, $\gamma^*(G) = 1/n$ for complete graphs, and $\gamma^*(G) = (n-1)/2n$ for rings with n odd.

4

The SER State Space

The state space of SER is the set Ω of all the acyclic orientations of G. In Ω, every orientation belongs to one basin of attraction, at whose core lies a period of orientations that acts as an attractor for all the orientations in that basin. The most important problem to be solved in connection with SER is the choice of the first acyclic orientation of a schedule according to the criterion of optimizing a measure of concurrency. This measure of concurrency is the rational number $\gamma(\omega)$ for orientation ω. The maximum value of this quantity over Ω is $\gamma^*(G)$.

Although there exist graphs for which $\gamma^*(G)$ is known analytically and can be computed easily as a function of n, the problem of determining its value is in general intractable, in the sense of NP-completeness [33]. The purpose of this chapter is to present data on the state space of SER that may in the future yield useful heuristics for the optimization of concurrency. Of course, like other efforts similar to the one that led to the creation of this book (e.g., [66]), the great complexity of the dynamical system's state space limits severely the degree of thoroughness with which to investigate its properties.

This chapter is divided into two sections. Section 4.1 contains a brief account of the two sources of combinatorial explosion with which one must deal when studying the properties of SER. One of these is the number of nonisomorphic graphs for fixed n, the other the number of acyclic orientations for fixed G.

In Section 4.2, statistics and correlations are presented for modestly-sized graphs. The aim is to provide a source of information on how the

properties of SER whose assessment is computationally demanding relate
to the properties of graphs and orientations that can be assessed efficiently.

4.1. Enumerating graphs and acyclic orientations

In very broad terms, there are essentially two types of combinatorial structures that ultimately determine the value of $\gamma^*(G)$ if we start with a fixed
value for n. One of them is G itself; the other is the set Ω of all the acyclic
orientations once G has been fixed. Being able to enumerate all the possible
variations of both structures, at least for small values of n, is one essential
first step toward understanding how those variations ultimately affect the
value of $\gamma^*(G)$.

Enumerating all the nonisomorphic graphs for fixed n is no simple task,
and the reason for this is immediately apparent from the data given in Table
4.1. In that table, we give, for $2 \leq n \leq 10$, the number of nonisomorphic
graphs in general, the number of nonisomorphic bipartite graphs, and the
number of nonisomorphic trees, all on n nodes and considering connected
graphs only. While for $n < 5$ the enumeration is trivial and can be done
easily by hand, already for $n = 6$ one must resort to a nontrivial formal
characterization of nonisomorphism to generate all graphs [45]. Beyond
that, the growth is so quick that enumeration by computer is the only
way to proceed. For this reason, it is no surprise that, between the first
enumeration of all nonisomorphic graphs for $n = 6$ and for $n = 10$ [18], a
span of nearly forty years exists. With the aid of computer technology, it
is now possible to go a little farther than $n = 10$ and still remain within
reasonable bounds on the time required for completion [51].

Once G is fixed, the generation of Ω is another source of exponential
growth, as we recall from our discussion of complete graphs, trees, and rings
in Section 3.1. We mention as a side remark that, curiously, even for graphs
in general there exists a principled way of assessing the number of acyclic
orientations of G, although it is impractical in most cases. The basis for
this is the *chromatic polynomial* of G [15], which we denote by π_G. For
$q \geq 0$, $\pi_G(q)$ gives the number of distinct ways in which the nodes of G can
be colored by a total of at most q colors. The least value of q for which
$\pi_G(q) > 0$ is of course equal to $\chi(G)$, the graph's chromatic number. The
remarkable result that relates this polynomial to the size of Ω is that the
number of acyclic orientations of G is given by the absolute value of the
chromatic polynomial of G evaluated at the negative unit [63, 65], that is,

$$|\Omega| = (-1)^n \pi_G(-1).$$

One recent approach to the generation of Ω proceeds as follows. For $0 \leq
i < n - 1$, let G_i denote the subgraph of G whose node set is $\{n_0, \ldots, n_i\}$.

Table 4.1. The number of nonisomorphic connected graphs

n	General	Bipartite	Trees
2	1	1	1
3	2	1	1
4	6	3	2
5	21	5	3
6	112	17	6
7	853	44	11
8	11,117	182	23
9	261,080	730	47
10	11,716,571	4,032	106

If all the acyclic orientations of G_i are known, then each of them can be extended to yield acyclic orientations of G_{i+1} by simply assigning directions to the edges between n_{i+1} and its neighbors from G_i in such a way as not to form directed cycles involving such nodes. Working upward from G_0, all the acyclic orientations of G can be generated within $O(e|\Omega|)$ time and $O(e)$ space. This generation never requires more than $O(ne)$ time per orientation [10].

Another recent approach is to proceed likewise, but to start from an empty subgraph of G and continue upward by adding one new edge at a time. Whenever an edge is added, itself and the edges between its end nodes and the nodes belonging to the already oriented portion of the graph must be oriented acyclically [61]. This approach requires $O(n|\Omega|)$ time and $O(n^2)$ space overall, but never more than $O(n^2 e)$ time per orientation. Neither approach is strictly better than the other in all aspects, so choosing between them must take into account which of the required computational resources is most constrained.

4.2. Some statistics and correlations

This section contains statistics and correlation data on several quantities that relate to the structure of G and to the SER schedules that originate from the acyclic orientations of G. All data are presented for classes of connected graphs with $2 \le n \le 10$, depending on the number of nonisomorphic graphs for fixed n and the number of acyclic orientations per graph. The techniques used to generate the graphs in each class and the acyclic orientations of each graph were those mentioned in Section 4.1 ([51] for graphs, [10] for acyclic orientations).

The graph classes that we consider are the following:

- $\mathcal{G}_n$: All connected graphs on n nodes.

- $\mathcal{B}_n$: All connected bipartite graphs on n nodes.

- $\mathcal{T}_n$: All trees on n nodes.

- $\mathcal{K}_n$: The complete graph on n nodes.

- $\mathcal{R}_n$: The ring on n nodes.

Several of the data to be presented are correlation data on two quantities relating to the same graph or the same acyclic orientation. If X and Y are these two quantities and we have z values for each of them, respectively $X_1, \ldots, X_z$ and $Y_1, \ldots, Y_z$, then the correlation indicator that we present is the so-called *correlation coefficient* [64], given by

$$\rho(X,Y) = \frac{\epsilon(XY) - \epsilon(X)\epsilon(Y)}{\sqrt{\left(\epsilon(X^2) - \epsilon^2(X)\right)\left(\epsilon(Y^2) - \epsilon^2(Y)\right)}}, \tag{4.1}$$

where ϵ denotes the average of the z values and ϵ^2 the square of that average. Clearly, $\rho(X,Y)$ is defined if and only if neither X nor Y is constant over its z values.

It follows from (4.1) that $-1 \leq \rho(X,Y) \leq 1$. In particular, if X and Y are linearly correlated to each other, that is, if there exists a nonzero real number a such that $Y_k = aX_k$ for $k = 1, \ldots, z$, then $\rho(X,Y) = 1$ if $a > 0$ and $\rho(X,Y) = -1$ if $a < 0$. Total uncorrelation between the two quantities is detected by $\rho(X,Y) = 0$.

The first group of correlations that we present relates quantities that can be measured efficiently on G to quantities that depend on the SER dynamics. By computing $\rho(X,Y)$ with X of the first type and Y of the second, we obtain data on how the structure of the undirected G ultimately affects the performance of SER, without regard to specific initial conditions. Each correlation is computed on all the graphs from a certain class $\mathcal{H}$, so z is the number of graphs in $\mathcal{H}$. In what follows, the list from which we select X is the one given next:

- e: The number of edges in G.

- $\delta_{\min}$: The minimum degree (number of neighbors) of a node in G.

- $\delta_{\max}$: The maximum degree of a node in G.

Table 4.2. $\rho(X, Y)$ for $Y = b$ on general graphs

Class	$X = e$	$X = \delta_{\min}$	$X = \delta_{\max}$	$X = D$	$X = g$
$\mathcal{G}_3$	1.000	1.000	—	−1.000	1.000
$\mathcal{G}_4$	0.984	0.968	0.333	−0.815	0.668
$\mathcal{G}_5$	0.954	0.923	0.435	−0.646	0.279
$\mathcal{G}_6$	0.908	0.873	0.441	−0.577	0.083
$\mathcal{G}_7$	0.866	0.807	0.466	−0.545	−0.025
$\mathcal{G}_8$	0.841	0.752	0.467	−0.548	−0.054

- D: The diameter of G, i.e., the maximum distance over all pairs of nodes.

- g: The girth of G, i.e., the number of nodes in the smallest undirected cycle in G.

As for Y, the first group of correlations employs selections from the following list:

- b: The number of basins of attraction in Ω.

- $\gamma_{\min}$: The minimum value of γ over all basins of attraction.

- $\gamma^*(G)$: By (3.4), the maximum value of γ over all basins of attraction.

The first group of correlation data appears in Tables 4.2 through 4.6, each with fixed Y and one column for each possible X. The first three tables have one row for each of $\mathcal{G}_3$ through $\mathcal{G}_8$, and the last two one row for each of $\mathcal{B}_4$ through $\mathcal{B}_9$. Of the five graph classes introduced earlier, those not contemplated in these tables have X or Y constant. Similarly for $Y = \gamma^*(G)$ on bipartite graphs and for entries missing from the tables.

Some entries in these tables do show strong correlations, but there is a tendency for great variation as n changes, sometimes even with changes in sign. Particularly useful correlations are those that remain constant or vary relatively slowly as n changes. This happens, for example, with $X = \delta_{\max}$ for general graphs, but not concomitantly with a very strong correlation.

Our second group of correlations refers to all the acyclic orientations of all the graphs within a certain class. As with the first group, the aim is to observe how quantities that depend on acyclic orientations and can be assessed efficiently relate to quantities that depend on the dynamics of

Table 4.3. $\rho(X, Y)$ for $Y = \gamma_{\min}$ on general graphs

Class	$X = e$	$X = \delta_{\min}$	$X = \delta_{\max}$	$X = D$	$X = g$
$\mathcal{G}_3$	-1.000	-1.000	—	1.000	-1.000
$\mathcal{G}_4$	-0.833	-0.777	-0.175	0.645	-0.960
$\mathcal{G}_5$	-0.736	-0.576	-0.234	0.581	-0.840
$\mathcal{G}_6$	-0.686	-0.515	-0.267	0.529	-0.674
$\mathcal{G}_7$	-0.638	-0.463	-0.274	0.528	-0.409
$\mathcal{G}_8$	-0.588	-0.460	-0.267	0.514	-0.156

Table 4.4. $\rho(X, Y)$ for $Y = \gamma^*(G)$ on general graphs

Class	$X = e$	$X = \delta_{\min}$	$X = \delta_{\max}$	$X = D$	$X = g$
$\mathcal{G}_3$	-1.000	-1.000	—	1.000	-1.000
$\mathcal{G}_4$	-0.880	-0.614	-0.680	0.714	-0.510
$\mathcal{G}_5$	-0.788	-0.505	-0.540	0.551	-0.392
$\mathcal{G}_6$	-0.713	-0.348	-0.599	0.480	-0.082
$\mathcal{G}_7$	-0.694	-0.336	-0.562	0.445	0.101
$\mathcal{G}_8$	-0.679	-0.303	-0.573	0.415	0.242

Table 4.5. $\rho(X, Y)$ for $Y = b$ on bipartite graphs

Class	$X = e$	$X = \delta_{\min}$	$X = \delta_{\max}$	$X = D$	$X = g$
$\mathcal{B}_4$	1.000	1.000	-0.500	-0.500	1.000
$\mathcal{B}_5$	0.986	0.943	0.000	-0.504	0.840
$\mathcal{B}_6$	0.925	0.914	-0.064	-0.548	0.495
$\mathcal{B}_7$	0.845	0.819	0.149	-0.496	0.319
$\mathcal{B}_8$	0.768	0.766	0.123	-0.441	0.146
$\mathcal{B}_9$	0.690	0.645	0.202	-0.388	0.084

Table 4.6. $\rho(X, Y)$ for $Y = \gamma_{\min}$ on bipartite graphs

Class	$X = e$	$X = \delta_{\min}$	$X = \delta_{\max}$	$X = D$	$X = g$
$\mathcal{B}_4$	-1.000	-1.000	0.500	0.500	-1.000
$\mathcal{B}_5$	-0.919	-0.612	0.000	0.327	-1.000
$\mathcal{B}_6$	-0.800	-0.601	0.181	0.341	-0.969
$\mathcal{B}_7$	-0.766	-0.397	-0.029	0.342	-0.958
$\mathcal{B}_8$	-0.693	-0.395	0.000	0.393	-0.888
$\mathcal{B}_9$	-0.660	-0.326	-0.067	0.416	-0.827

SER. Correlation coefficients $\rho(X, Y)$ with X of the first type and Y of the second carry information on how certain properties of acyclic orientations affect the behavior of the SER schedules starting at those orientations. Each correlation is computed over all the acyclic orientations of all graphs in a certain class $\mathcal{H}$, so z is the sum of $|\Omega|$ for all graphs in $\mathcal{H}$. For this second group of correlations, X is selected from the following list:

- $\delta_{\max}^+$: The maximum in-degree according to the orientation. The in-degree of a node is the number of neighbors from which edges are oriented toward the node.

- $\delta_{\max}^-$: The maximum out-degree according to the orientation. The out-degree of a node is the number of neighbors toward which edges are oriented from the node.

- λ: The length of the sink decomposition according to the orientation. The length of the longest directed path according to the orientation is $\lambda - 1$.

- σ^+: The number of sinks according to the orientation.

- σ^-: The number of sources according to the orientation.

The list from which Y is selected for the second group of correlations is the following:

- τ: The number of transient orientations on the schedule starting at the orientation.

- φ: The fraction $\gamma/\gamma^*(G)$ of optimal concurrency achieved from the orientation.

Table 4.7. $\rho(X, Y)$ for $Y = \tau$ on general graphs

Class	$X = \delta^+_{\max}$	$X = \delta^-_{\max}$	$X = \lambda$	$X = \sigma^+$	$X = \sigma^-$
$\mathcal{G}_3$	−0.764	−0.764	0.250	−0.167	−0.167
$\mathcal{G}_4$	−0.675	−0.609	−0.124	0.142	−0.052
$\mathcal{G}_5$	−0.568	−0.501	−0.111	0.216	−0.047
$\mathcal{G}_6$	−0.514	−0.433	−0.134	0.255	−0.054
$\mathcal{G}_7$	−0.460	−0.374	−0.126	0.265	−0.070
$\mathcal{G}_8$	−0.408	−0.318	−0.112	0.265	−0.091

Table 4.8. $\rho(X, Y)$ for $Y = \varphi$ on general graphs

Class	$X = \delta^+_{\max}$	$X = \delta^-_{\max}$	$X = \lambda$	$X = \sigma^+$	$X = \sigma^-$
$\mathcal{G}_4$	0.110	0.110	−0.346	0.214	0.214
$\mathcal{G}_5$	0.106	0.106	−0.355	0.229	0.229
$\mathcal{G}_6$	0.128	0.128	−0.389	0.229	0.229
$\mathcal{G}_7$	0.125	0.125	−0.440	0.226	0.226
$\mathcal{G}_8$	0.124	0.124	−0.497	0.225	0.225

The second group of correlations appears in Tables 4.7 through 4.13, each with fixed Y and one column for each possible X. Table 4.7 has one row for each of $\mathcal{G}_3$ through $\mathcal{G}_8$, Table 4.8 for $\mathcal{G}_4$ through $\mathcal{G}_8$, Table 4.9 for $\mathcal{B}_3$ through $\mathcal{B}_9$, Table 4.10 for $\mathcal{B}_4$ through $\mathcal{B}_9$, Table 4.11 for $\mathcal{T}_3$ through $\mathcal{T}_{10}$, and Tables 4.12 and 4.13 for $\mathcal{R}_4$ through $\mathcal{R}_{10}$. As before, graph classes not contemplated in these tables have X or Y constant, the same holding for $Y = \varphi$ on trees and for columns missing from Tables 4.12 and 4.13.

As with the first group of correlations, interesting information from these tables comes from correlations that are significantly distant from zero and relatively stable for entire columns. For example, it appears that shorter sink decompositions tend to lead to higher fractions of the optimal concurrency, the same holding, although not as strongly, for larger numbers of sinks and sources.

The third and final set of data that we present is related to the sensitivity of SER to the choice of the initial acyclic orientation. This sensitivity

Table 4.9. $\rho(X, Y)$ for $Y = \tau$ on bipartite graphs

Class	$X = \delta_{max}^+$	$X = \delta_{max}^-$	$X = \lambda$	$X = \sigma^+$	$X = \sigma^-$
B_3	−0.577	−0.577	1.000	−0.577	−0.577
B_4	−0.609	−0.417	0.089	−0.029	−0.203
B_5	−0.501	−0.400	0.087	0.070	−0.091
B_6	−0.489	−0.380	−0.088	0.181	−0.018
B_7	−0.469	−0.359	−0.068	0.201	−0.026
B_8	−0.461	−0.349	−0.103	0.230	−0.013
B_9	−0.426	−0.313	−0.086	0.221	−0.034

Table 4.10. $\rho(X, Y)$ for $Y = \varphi$ on bipartite graphs

Class	$X = \delta_{max}^+$	$X = \delta_{max}^-$	$X = \lambda$	$X = \sigma^+$	$X = \sigma^-$
B_4	−0.284	−0.284	−0.728	0.404	0.404
B_5	−0.249	−0.249	−0.723	0.426	0.426
B_6	−0.296	−0.296	−0.748	0.460	0.460
B_7	−0.266	−0.266	−0.721	0.428	0.428
B_8	−0.273	−0.273	−0.723	0.426	0.426
B_9	−0.244	−0.244	−0.724	0.397	0.397

is crucial for the design of heuristics to select the initial orientation, because it indicates the deviation that may occur from optimal concurrency as a result of slightly imprecise design choices in the beginning. It is, in this sense, related to the notion of deterministic chaos in dynamical systems [59].

We present these sensitivity data as eight different measures, to be explained shortly and denoted by $\alpha_1, \ldots, \alpha_8$. Each measure indicates an average or bound related to the effect of reversing the orientation of one single edge to yield another acyclic orientation. Like other data presented so far in this chapter, these measures refer to all acyclic orientations of all graphs in a certain class $\mathcal{H}$.

Some of the measures are averages, of which we consider two types. *Overall averages* are taken over all graphs in $\mathcal{H}$, all acyclic orientations

Table 4.11. $\rho(X,Y)$ for $Y = \tau$ on trees

Class	$X = \delta^+_{\max}$	$X = \delta^-_{\max}$	$X = \lambda$	$X = \sigma^+$	$X = \sigma^-$
T_3	−0.577	−0.577	1.000	−0.577	−0.577
T_4	−0.580	−0.290	0.885	−0.290	−0.580
T_5	−0.413	−0.242	0.797	−0.258	−0.453
T_6	−0.426	−0.247	0.706	−0.212	−0.447
T_7	−0.375	−0.262	0.652	−0.218	−0.394
T_8	−0.379	−0.273	0.586	−0.200	−0.380
T_9	−0.347	−0.268	0.547	−0.196	−0.351
T_{10}	−0.336	−0.265	0.501	−0.186	−0.335

Table 4.12. $\rho(X,Y)$ for $Y = \tau$ on rings

Class	$X = \lambda$	$X = \sigma^+$	$X = \sigma^-$
R_4	−0.372	−0.258	−0.258
R_5	0.000	−0.500	−0.500
R_6	−0.278	−0.187	−0.187
R_7	−0.100	−0.325	−0.325
R_8	−0.207	−0.153	−0.153
R_9	−0.102	−0.238	−0.238
R_{10}	−0.165	−0.128	−0.128

of each graph, and all single-edge reversals that lead to another acyclic orientation. By contrast, *per-orientation averages* are computed in two phases. First the quantity of interest is averaged for a fixed acyclic orientation over all single-edge reversals that yield another acyclic orientation. Then these averages are themselves averaged over all graphs in $\mathcal{H}$ and all acyclic orientations of each graph from which at least one other acyclic orientation can be derived by means of a single-edge reversal.

Other measures are bounds (minima or maxima), being evaluated over all graphs in $\mathcal{H}$, all acyclic orientations of each graph, and all single-edge reversals that result in another acyclic orientation.

Table 4.13. $\rho(X, Y)$ for $Y = \varphi$ on rings

Class	$X = \lambda$	$X = \sigma^+$	$X = \sigma^-$
$\mathcal{R}_4$	-0.906	0.471	0.471
$\mathcal{R}_5$	-0.866	0.500	0.500
$\mathcal{R}_6$	-0.825	0.474	0.474
$\mathcal{R}_7$	-0.803	0.488	0.488
$\mathcal{R}_8$	-0.763	0.454	0.454
$\mathcal{R}_9$	-0.744	0.454	0.454
$\mathcal{R}_{10}$	-0.705	0.422	0.422

The following is how α_1 through α_8 are defined:

- α_1: The overall average rate of basin change. The quantity being averaged for each single-edge reversal that preserves acyclicity is 1 for reversals that lead to another basin of attraction, 0 for the others.

- α_2: The per-orientation average rate of basin change. The quantity being averaged is the same as in α_1.

- α_3: The overall average rate of concurrency change. The quantity being averaged for each single-edge reversal that preserves acyclicity is 1 for reversals that lead to a basin of different concurrency, 0 for the others.

- α_4: The per-orientation average rate of concurrency change. The quantity being averaged is the same as in α_3.

- α_5: The overall average concurrency change. The quantity being averaged for each single-edge reversal that preserves acyclicity is $|\gamma(\omega) - \gamma(\omega')|$, where ω and ω' are the two orientations involved in the single-edge reversal.

- α_6: The per-orientation average concurrency change. The quantity being averaged is the same as in α_5.

- α_7: The minimum change in concurrency, as given for α_5.

- α_8: The maximum change in concurrency, as given for α_5.

We present these eight measures in Tables 4.14 through 4.21 for $\mathcal{G}_2$ through $\mathcal{G}_8$ (one row of Tables 4.14 and 4.15 for each), $\mathcal{B}_2$ through $\mathcal{B}_9$ (Tables 4.16 and 4.17), $\mathcal{K}_2$ through $\mathcal{K}_9$ (Tables 4.18 and 4.19), and $\mathcal{R}_3$ through $\mathcal{R}_{10}$

Table 4.14. α_1 through α_4 on general graphs

Class	α_1	α_2	α_3	α_4
$\mathcal{G}_2$	0.000	0.000	0.000	0.000
$\mathcal{G}_3$	0.600	0.600	0.000	0.000
$\mathcal{G}_4$	0.769	0.762	0.369	0.349
$\mathcal{G}_5$	0.880	0.874	0.443	0.429
$\mathcal{G}_6$	0.931	0.928	0.546	0.536
$\mathcal{G}_7$	0.960	0.959	0.586	0.579
$\mathcal{G}_8$	0.978	0.977	0.611	0.607

Table 4.15. α_5 through α_8 on general graphs

Class	α_5	α_6	α_7	α_8
$\mathcal{G}_2$	0.000	0.000	0.000	0.000
$\mathcal{G}_3$	0.000	0.000	0.000	0.000
$\mathcal{G}_4$	0.062	0.057	0.000	0.250
$\mathcal{G}_5$	0.046	0.043	0.000	0.250
$\mathcal{G}_6$	0.040	0.039	0.000	0.250
$\mathcal{G}_7$	0.032	0.031	0.000	0.250
$\mathcal{G}_8$	0.026	0.025	0.000	0.250

(Tables 4.20 and 4.21). Trees are altogether absent, because the SER state space for trees has one single basin of attraction, so $\alpha_1, \ldots, \alpha_8$ are zero throughout. Tables 4.14, 4.16, 4.18, and 4.20 refer to measures $\alpha_1, \ldots, \alpha_4$, while Tables 4.15, 4.17, 4.19, and 4.21 refer to measures $\alpha_5, \ldots, \alpha_8$.

Recalling that average rates must be between 0 and 1, and average changes in concurrency between 0 and 0.5, what most of these tables tell us is that there is indeed considerable sensitivity to the small perturbations caused by single-edge reversals. Extreme cases of interest are those of complete graphs (single-edge reversals always cause basin changes, but, as expected, no changes in concurrency) and of rings (single-edge reversals always cause basin changes and, if n is even, concurrency changes as well).

Table 4.16. α_1 through α_4 on bipartite graphs

Class	α_1	α_2	α_3	α_4
$\mathcal{B}_2$	0.000	0.000	0.000	0.000
$\mathcal{B}_3$	0.000	0.000	0.000	0.000
$\mathcal{B}_4$	0.500	0.467	0.500	0.467
$\mathcal{B}_5$	0.586	0.554	0.406	0.377
$\mathcal{B}_6$	0.783	0.761	0.509	0.491
$\mathcal{B}_7$	0.818	0.801	0.458	0.444
$\mathcal{B}_8$	0.885	0.875	0.469	0.459
$\mathcal{B}_9$	0.907	0.899	0.449	0.441

Table 4.17. α_5 through α_8 on bipartite graphs

Class	α_5	α_6	α_7	α_8
$\mathcal{B}_2$	0.000	0.000	0.000	0.000
$\mathcal{B}_3$	0.000	0.000	0.000	0.000
$\mathcal{B}_4$	0.125	0.117	0.000	0.250
$\mathcal{B}_5$	0.102	0.094	0.000	0.250
$\mathcal{B}_6$	0.099	0.094	0.000	0.250
$\mathcal{B}_7$	0.078	0.075	0.000	0.250
$\mathcal{B}_8$	0.060	0.058	0.000	0.250
$\mathcal{B}_9$	0.047	0.046	0.000	0.250

Table 4.18. α_1 through α_4 on complete graphs

Class	α_1	α_2	α_3	α_4
$\mathcal{K}_2$	0.000	0.000	0.000	0.000
$\mathcal{K}_3$	1.000	1.000	0.000	0.000
$\mathcal{K}_4$	1.000	1.000	0.000	0.000
$\mathcal{K}_5$	1.000	1.000	0.000	0.000
$\mathcal{K}_6$	1.000	1.000	0.000	0.000
$\mathcal{K}_7$	1.000	1.000	0.000	0.000
$\mathcal{K}_8$	1.000	1.000	0.000	0.000
$\mathcal{K}_9$	1.000	1.000	0.000	0.000

Table 4.19. α_5 through α_8 on complete graphs

Class	α_5	α_6	α_7	α_8
$\mathcal{K}_2$	0.000	0.000	0.000	0.000
$\mathcal{K}_3$	0.000	0.000	0.000	0.000
$\mathcal{K}_4$	0.000	0.000	0.000	0.000
$\mathcal{K}_5$	0.000	0.000	0.000	0.000
$\mathcal{K}_6$	0.000	0.000	0.000	0.000
$\mathcal{K}_7$	0.000	0.000	0.000	0.000
$\mathcal{K}_8$	0.000	0.000	0.000	0.000
$\mathcal{K}_9$	0.000	0.000	0.000	0.000

Table 4.20. α_1 through α_4 on rings

Class	α_1	α_2	α_3	α_4
$\mathcal{R}_3$	1.000	1.000	0.000	0.000
$\mathcal{R}_4$	1.000	1.000	1.000	1.000
$\mathcal{R}_5$	1.000	1.000	0.571	0.600
$\mathcal{R}_6$	1.000	1.000	1.000	1.000
$\mathcal{R}_7$	1.000	1.000	0.677	0.683
$\mathcal{R}_8$	1.000	1.000	1.000	1.000
$\mathcal{R}_9$	1.000	1.000	0.724	0.725
$\mathcal{R}_{10}$	1.000	1.000	1.000	1.000

Table 4.21. α_5 through α_8 on rings

Class	α_5	α_6	α_7	α_8
$\mathcal{R}_3$	0.000	0.000	0.000	0.000
$\mathcal{R}_4$	0.250	0.250	0.250	0.250
$\mathcal{R}_5$	0.114	0.120	0.000	0.200
$\mathcal{R}_6$	0.167	0.167	0.167	0.167
$\mathcal{R}_7$	0.097	0.098	0.000	0.143
$\mathcal{R}_8$	0.125	0.125	0.125	0.125
$\mathcal{R}_9$	0.080	0.081	0.000	0.111
$\mathcal{R}_{10}$	0.100	0.100	0.100	0.100

II

The Atlas

This second part of the book is an atlas of SER basins of attraction. For all graphs in selected classes, Chapters 5 through 7 present all the basins of attraction. Generation of all graphs within each class and of all the acyclic orientations of each graph was based on the techniques mentioned in Section 4.1 ([51] for graphs, [10] for acyclic orientations).

Two main criteria were adopted when selecting the graph classes to appear in the atlas. First, such a class should contain enough variation of SER behavior in order to provide a rich insight into the dynamics of SER. The second criterion, just as fundamental as the first, was that a graph class to appear in the atlas should comprise a number of graphs, and each graph a number of orientations, compatible with the various space constraints for the book. These criteria led to the choice of classes $\mathcal{G}_6$ (Chapter 5), $\mathcal{T}_7$ (Chapter 6), and $\mathcal{R}_3$ through $\mathcal{R}_8$ (Chapter 7).

Each graph appears in the atlas as a series of entries, one containing some of the graph's characteristics (a graph entry), and one for each group of the graph's basins of attraction having the same values for m and p (a basin entry). The graph entry is indicated by the symbol ▶, followed by the values of n and e for that graph, as well as the range of δ (the nodes' degrees). The latter is given as either $\delta \in \{\delta_{min}, \ldots, \delta_{max}\}$, $\delta \in \{\delta_{min}, \delta_{max}\}$, or $\delta = \delta_{min}$, respectively for $\delta_{min} < \delta_{max} - 1$, $\delta_{min} = \delta_{max} - 1$, and $\delta_{min} = \delta_{max}$. A basin entry is indicated by the symbol ▷, followed by the corresponding values of γ, m, and p. For each graph, basin entries appear in order of nondecreasing γ, and, for constant γ, in order of decreasing p.

A graph entry contains a picture of the undirected graph. In this picture, nodes are displayed on a circle counterclockwise in increasing order of subscripts, and edges are displayed as chords of the circle. Nodes are identified by their subscripts, ranging from 0 through $n - 1$. For example, the identification that appears by node n_2 is 2.

A basin entry contains, for each basin of attraction, a picture of the entire basin. In this picture, each orientation is shown in a rectangular box, and boxes are connected to each other to depict the SER schedules within the corresponding basin. All schedules are displayed from left to right while on transient orientations, then counterclockwise within the period.

Each acyclic orientation is displayed as a total order of node subscripts, for example 2350164 for a 7-node graph. Of course, acyclic orientations induce partial orders, not total orders, so the interpretation of this sequence of subscripts is dependent upon the structure of G. If, for example, (n_1, n_5) is an edge of G, then, in the orientation represented by 2350164, this edge is oriented from n_5 to n_1, that is, the total order used to represent the orientation is consistent with the partial order induced by the orientation. Note that such a total order represents the orientation unambiguously given G, although not uniquely.

Next are two pairs of entries, provided for the sake of illustration, each containing one graph entry and one basin entry. The first pair's basin entry contains the basin appearing in Figure 1.2, the second pair's the basin of Figure 3.4. Although in those figures nodes are not identified, the correspondence should be clear.

▸ $n = 5, e = 5, \delta = 2$

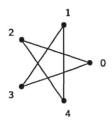

▷ $\gamma = 0.4$ $(m = 2, p = 5)$

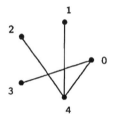

▶ $n = 5$, $e = 4$, $\delta \in \{1, \ldots, 3\}$

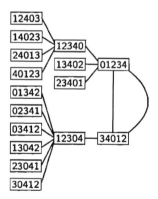

▷ $\gamma = 0.5$ $(m = 1, p = 2)$

All graphs displayed in the atlas appear in the Index under "graph in atlas," and are listed according to the values of n and e and the range of δ. The entry "graph in atlas, $n = 6$, $e = 10$, $\delta \in \{3, \ldots, 5\}$," for example, points to all pages on which a graph entry can be found with $n = 6$, $e = 10$, and $\delta \in \{3, \ldots, 5\}$.

5

All Graphs on Six Nodes

This chapter is the atlas of SER basins of attraction for graphs in $\mathcal{G}_6$. It contains entries for all the 112 connected graphs with $n = 6$.

▶ $n = 6$, $e = 5$, $\delta \in \{1,\ldots,5\}$

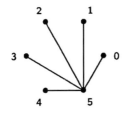

▷ $\gamma = 0.5$ $(m = 1, p = 2)$

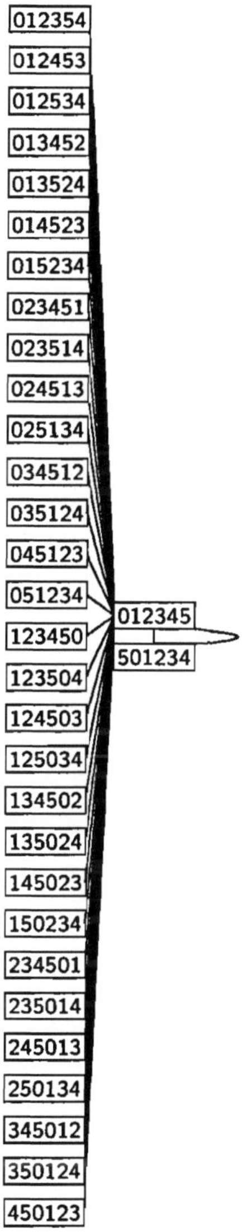

012354
012453
012534
013452
013524
014523
015234
023451
023514
024513
025134
034512
035124
045123
051234
123450
123504
124503
125034
134502
135024
145023
150234
234501
235014
245013
250134
345012
350124
450123

012345
501234

▶ $n = 6$, $e = 5$, $\delta \in \{1, \ldots, 4\}$

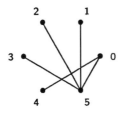

▷ $\gamma = 0.5$ $(m = 1, p = 2)$

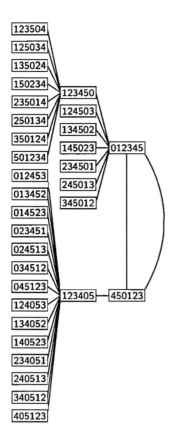

▶ $n = 6$, $e = 6$, $\delta \in \{1, \ldots, 5\}$

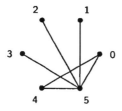

▷ $\gamma = 0.333333 \; (m = 1, p = 3)$

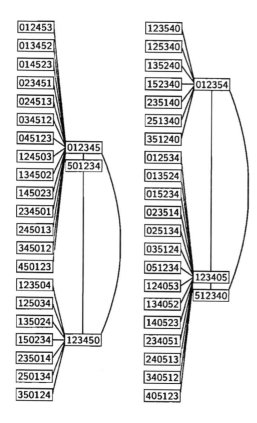

▸ $n = 6$, $e = 5$, $\delta \in \{1, \ldots, 3\}$

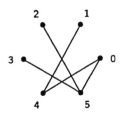

▷ $\gamma = 0.5$ $(m = 1,\ p = 2)$

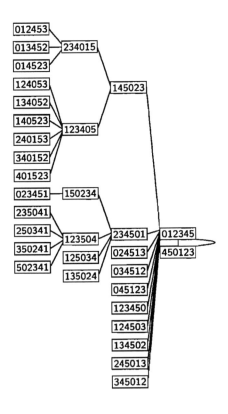

▶ $n = 6$, $e = 5$, $\delta \in \{1, \ldots, 3\}$

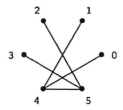

▷ $\gamma = 0.5$ $(m = 1,\ p = 2)$

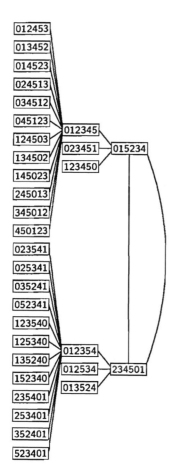

▶ $n = 6$, $e = 6$, $\delta \in \{1, \dots, 4\}$

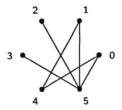

▷ $\gamma = 0.25$ $(m = 1,\ p = 4)$

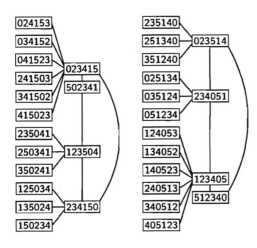

▷ $\gamma = 0.5\ (m = 1,\ p = 2)$

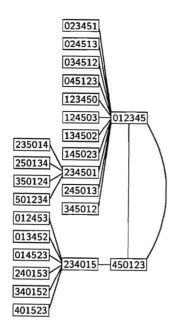

▶ $n = 6,\ e = 6,\ \delta \in \{1, \ldots, 4\}$

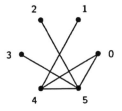

▷ $\gamma = 0.333333\ (m = 1,\ p = 3)$

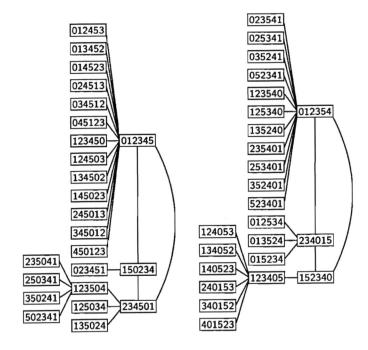

▶ $n = 6$, $e = 7$, $\delta \in \{1, \dots, 5\}$

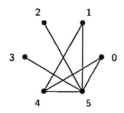

▷ $\gamma = 0.25$ $(m = 1, p = 4)$

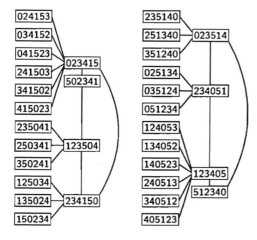

▷ $\gamma = 0.333333$ $(m = 1, p = 3)$

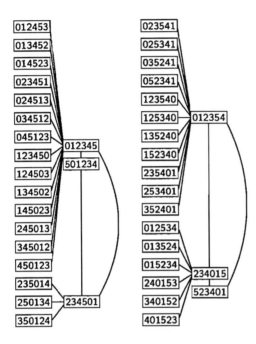

▶ $n = 6, e = 6, \delta \in \{1, \dots, 3\}$

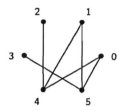

▷ $\gamma = 0.25 \ (m = 1, p = 4)$

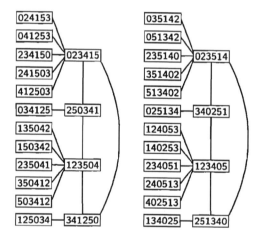

▷ $\gamma = 0.5$ $(m = 1, p = 2)$

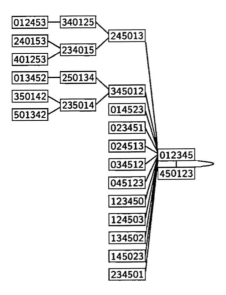

▸ $n = 6$, $e = 7$, $\delta \in \{1, \ldots, 4\}$

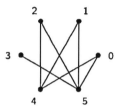

▷ $\gamma = 0.25$ $(m = 1, p = 4)$

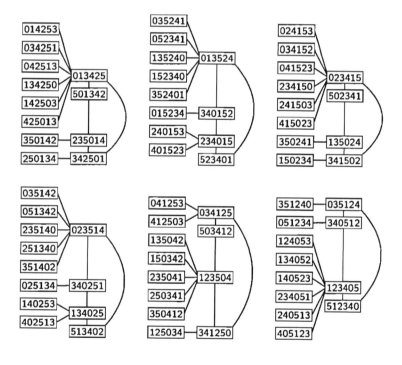

▷ $\gamma = 0.5 \ (m = 1, \ p = 2)$

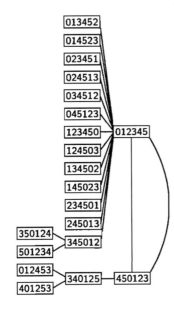

▶ $n = 6$, $e = 7$, $\delta \in \{1, \ldots, 4\}$

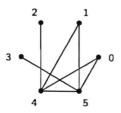

▷ $\gamma = 0.25$ $(m = 1, p = 4)$

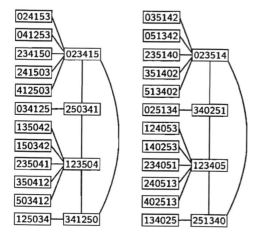

▷ $\gamma = 0.333333$ $(m = 1, p = 3)$

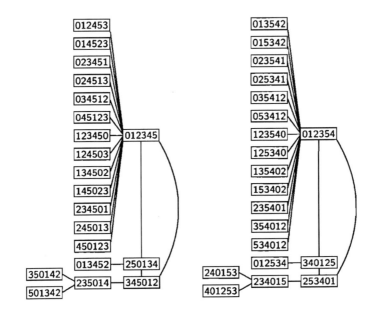

▶ $n = 6$, $e = 8$, $\delta \in \{1, \ldots, 5\}$

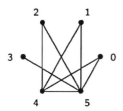

▷ $\gamma = 0.25$ $(m = 1, p = 4)$

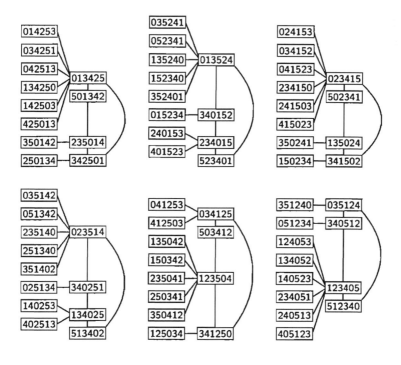

▷ $\gamma = 0.333333$ $(m = 1, p = 3)$

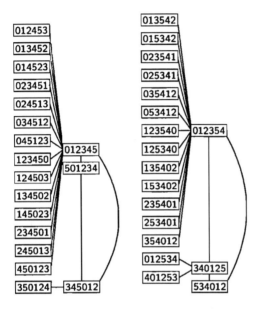

▶ $n = 6$, $e = 8$, $\delta \in \{2, \ldots, 4\}$

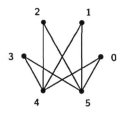

▷ $\gamma = 0.25$ $(m = 1, p = 4)$

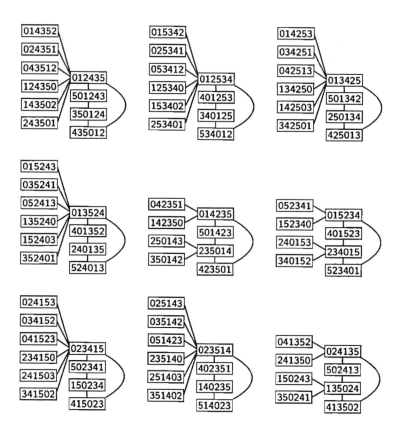

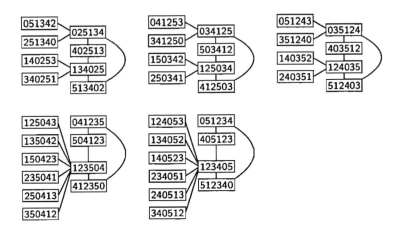

▷ $\gamma = 0.5\ (m = 1,\ p = 2)$

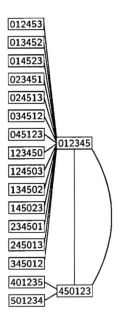

▶ $n = 6$, $e = 9$, $\delta \in \{2, \ldots, 5\}$

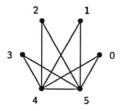

▷ $\gamma = 0.25$ ($m = 1$, $p = 4$)

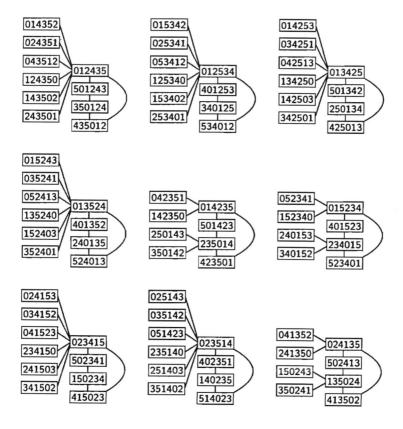

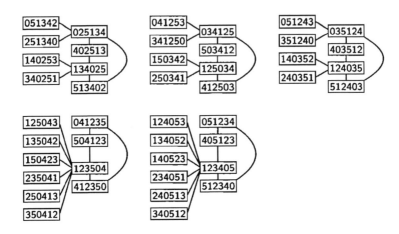

$\triangleright \quad \gamma = 0.333333 \; (m = 1, \, p = 3)$

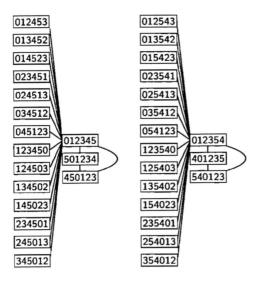

▶ $n = 6,\ e = 5,\ \delta \in \{1, \ldots, 3\}$

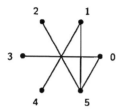

▷ $\gamma = 0.5\ (m = 1,\ p = 2)$

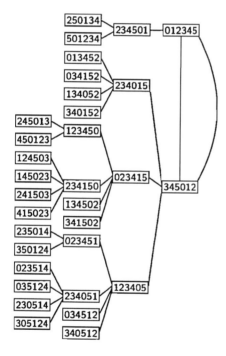

▶ $n = 6,\ e = 6,\ \delta \in \{1, \ldots, 4\}$

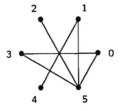

▷ $\gamma = 0.333333\ (m = 1,\ p = 3)$

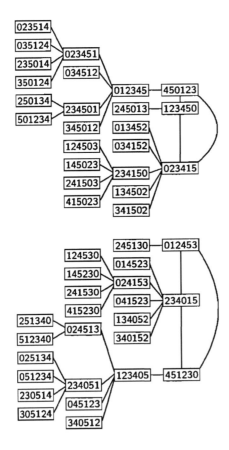

▸ $n = 6$, $e = 7$, $\delta \in \{1, \ldots, 5\}$

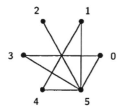

▷ $\gamma = 0.333333$ $(m = 1, p = 3)$

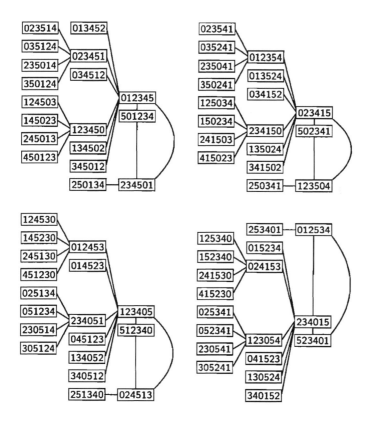

▶ $n = 6$, $e = 6$, $\delta \in \{1, \ldots, 3\}$

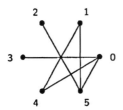

▷ $\gamma = 0.25$ $(m = 1, p = 4)$

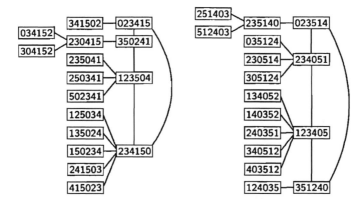

▷ $\gamma = 0.5 \ (m = 1, \ p = 2)$

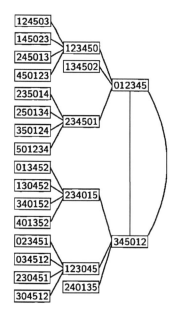

▶ $n = 6$, $e = 6$, $\delta \in \{1, \ldots, 3\}$

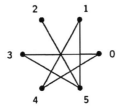

▷ $\gamma = 0.2$ ($m = 1$, $p = 5$)

▷ $\gamma = 0.4$ ($m = 2$, $p = 5$)

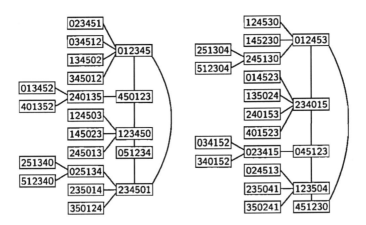

▶ $n = 6,\ e = 6,\ \delta \in \{1, \ldots, 3\}$

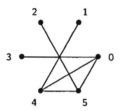

▷ $\gamma = 0.333333\ (m = 1,\ p = 3)$

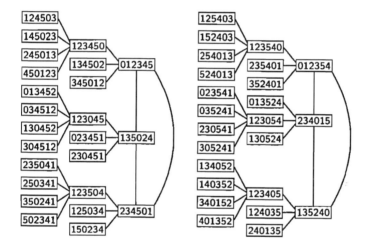

▶ $n = 6, e = 7, \delta \in \{1, \ldots, 4\}$

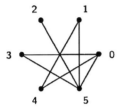

▷ $\gamma = 0.2 \; (m = 1, p = 5)$

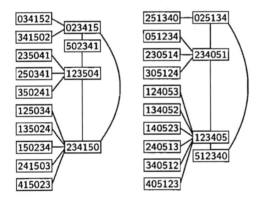

▷ $\gamma = 0.25 \; (m = 1, p = 4)$

▷ $\gamma = 0.333333$ $(m = 1,\, p = 3)$

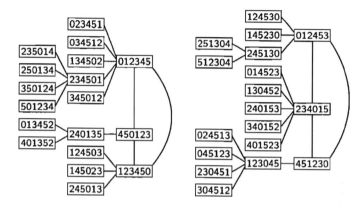

▶ $n = 6,\ e = 7,\ \delta \in \{1, \ldots, 4\}$

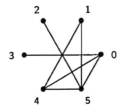

▷ $\gamma = 0.25\ (m = 1,\ p = 4)$

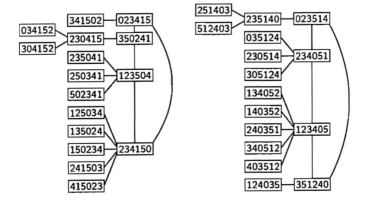

▷ $\gamma = 0.333333 \ (m = 1, \ p = 3)$

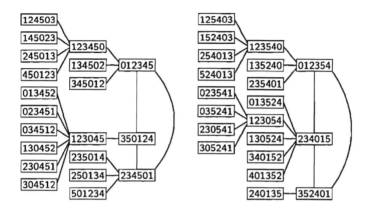

▶ $n = 6,\ e = 8,\ \delta \in \{1, \ldots, 5\}$

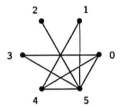

▷ $\gamma = 0.2\ (m = 1,\ p = 5)$

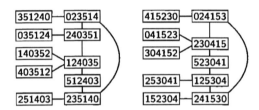

▷ $\gamma = 0.25\ (m = 1,\ p = 4)$

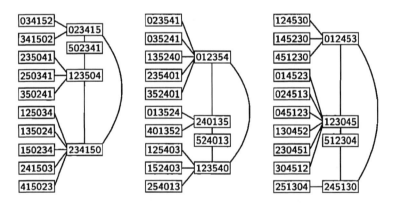

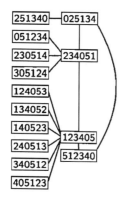

$\triangleright \quad \gamma = 0.333333 \ (m = 1, \ p = 3)$

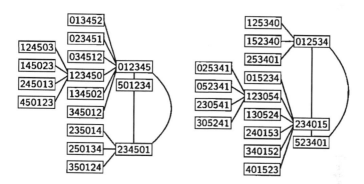

▶ $n = 6,\ e = 5,\ \delta \in \{1, 2\}$

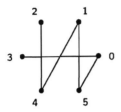

▷ $\gamma = 0.5\ (m = 1,\ p = 2)$

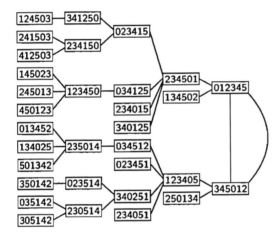

▶ $n = 6$, $e = 6$, $\delta \in \{1, \ldots, 3\}$

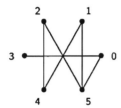

▷ $\gamma = 0.25$ $(m = 1, p = 4)$

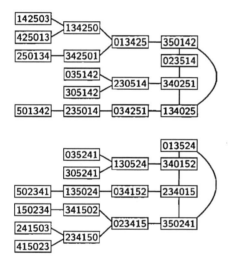

▷ $\gamma = 0.5 \; (m = 1, \, p = 2)$

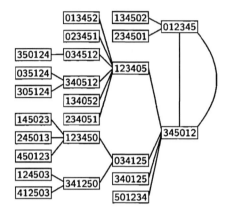

▶ $n = 6$, $e = 6$, $\delta \in \{1, \ldots, 3\}$

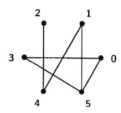

▷ $\gamma = 0.333333$ $(m = 1, p = 3)$

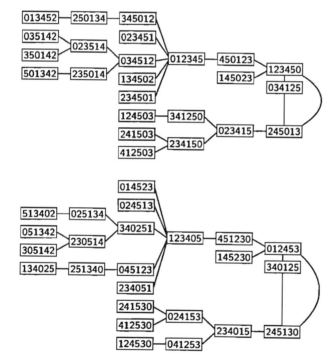

▶ $n = 6$, $e = 6$, $\delta \in \{1,\ldots,3\}$

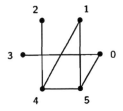

▷ $\gamma = 0.333333$ $(m = 1, p = 3)$

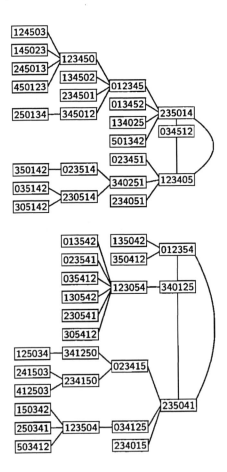

▶ $n = 6$, $e = 6$, $\delta \in \{1, \ldots, 3\}$

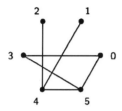

▷ $\gamma = 0.333333$ $(m = 1, p = 3)$

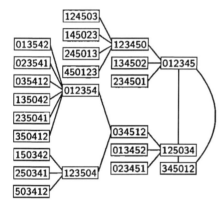

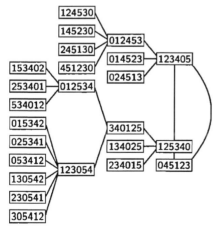

▶ $n = 6$, $e = 7$, $\delta \in \{2, \ldots, 4\}$

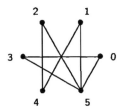

▷ $\gamma = 0.25$ $(m = 1, p = 4)$

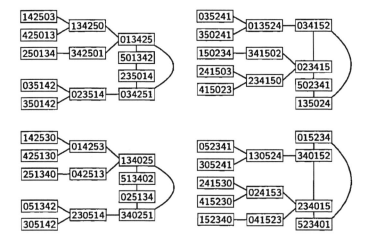

▷ $\gamma = 0.333333$ $(m = 1, p = 3)$

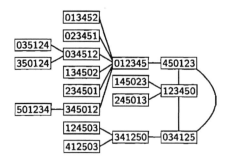

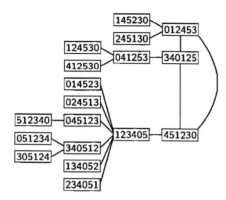

▶ $n = 6, e = 7, \delta \in \{1, \ldots, 4\}$

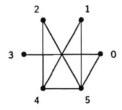

▷ $\gamma = 0.25$ $(m = 1, p = 4)$

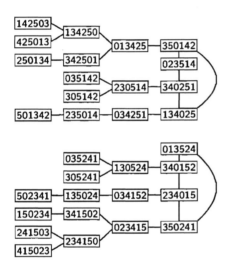

▷ $\gamma = 0.333333$ $(m = 1, p = 3)$

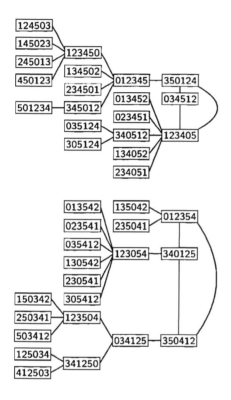

▶ $n = 6, e = 7, \delta \in \{1, \ldots, 4\}$

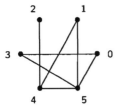

▷ $\gamma = 0.333333 \ (m = 1, p = 3)$

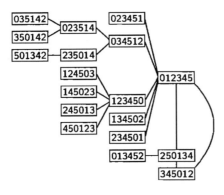

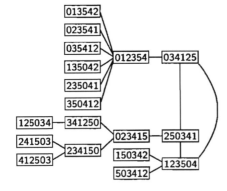

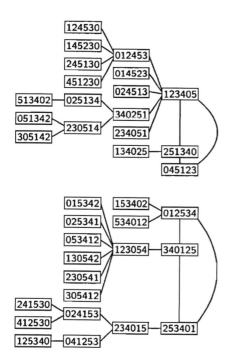

▶ $n = 6,\ e = 8,\ \delta \in \{2, \ldots, 5\}$

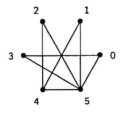

▷ $\gamma = 0.25\ (m = 1,\ p = 4)$

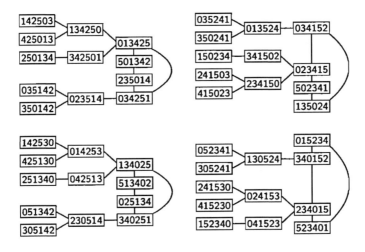

▷ $\gamma = 0.333333 \ (m = 1, \ p = 3)$

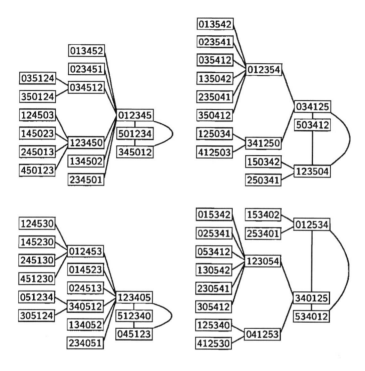

▶ $n = 6,\ e = 7,\ \delta \in \{1, \ldots, 4\}$

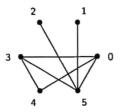

▷ $\gamma = 0.25\ (m = 1,\ p = 4)$

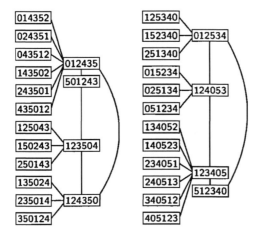

▷ $\gamma = 0.333333\ (m = 1,\ p = 3)$

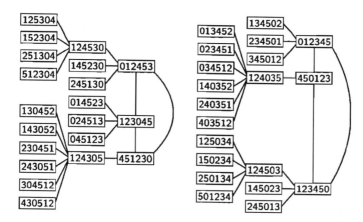

▸ $n = 6$, $e = 8$, $\delta \in \{1, \ldots, 5\}$

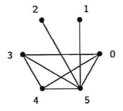

▷ $\gamma = 0.25$ $(m = 1, p = 4)$

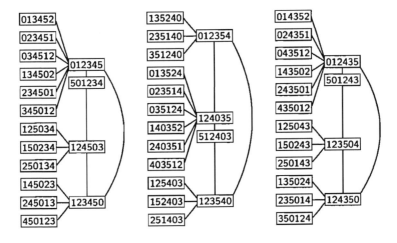

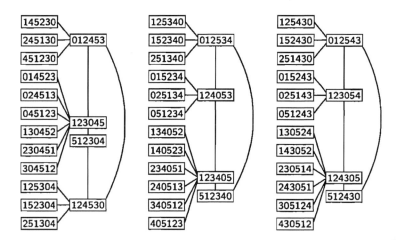

▶ $n = 6$, $e = 7$, $\delta \in \{2,3\}$

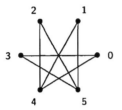

▷ $\gamma = 0.2$ ($m = 1$, $p = 5$)

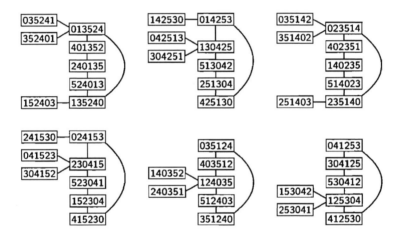

▷ $\gamma = 0.25$ ($m = 1$, $p = 4$)

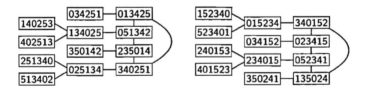

▷ $\gamma = 0.4 \ (m = 2, \ p = 5)$

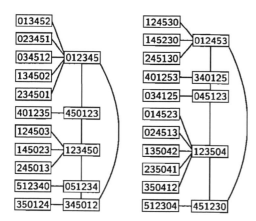

► $n = 6$, $e = 8$, $\delta \in \{2, \ldots, 4\}$

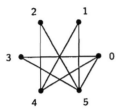

▷ $\gamma = 0.2$ $(m = 1, p = 5)$

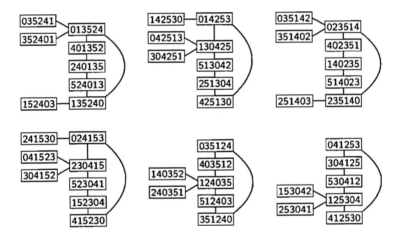

▷ $\gamma = 0.25$ $(m = 1, p = 4)$

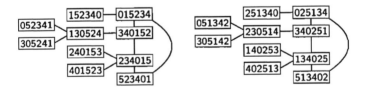

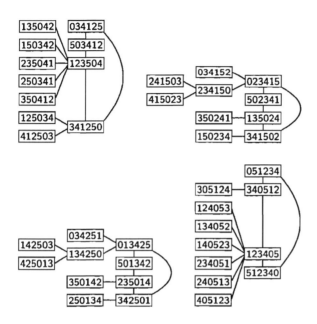

$\triangleright \quad \gamma = 0.333333 \ (m = 1, \ p = 3)$

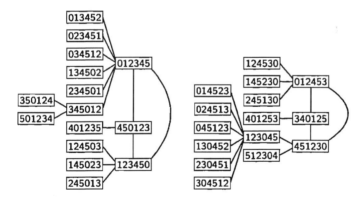

▶ $n = 6,\ e = 8,\ \delta \in \{1, \ldots, 4\}$

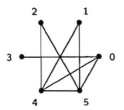

▷ $\gamma = 0.25\ (m = 1,\ p = 4)$

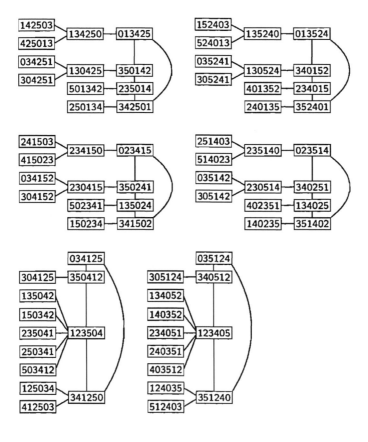

▷ $\gamma = 0.333333$ $(m = 1,\, p = 3)$

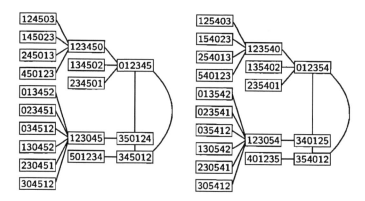

▶ $n = 6$, $e = 8$, $\delta \in \{1, \ldots, 4\}$

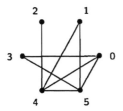

▷ $\gamma = 0.2$ $(m = 1, p = 5)$

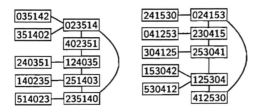

▷ $\gamma = 0.25$ $(m = 1, p = 4)$

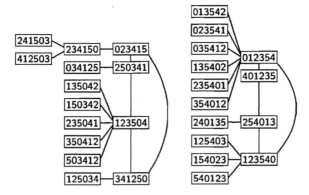

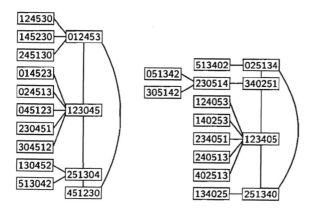

▷ $\gamma = 0.333333$ $(m = 1, p = 3)$

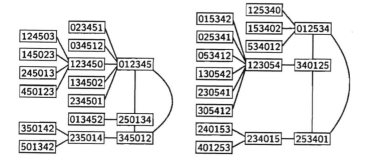

▶ $n = 6, \ e = 8, \ \delta \in \{2, \dots, 4\}$

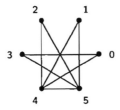

▷ $\gamma = 0.2 \ (m = 1, \ p = 5)$

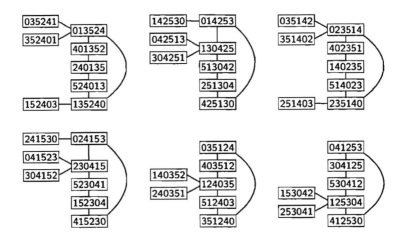

▷ $\gamma = 0.25\ (m = 1,\ p = 4)$

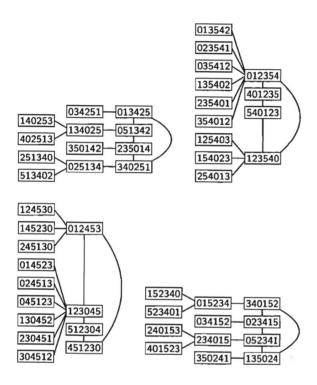

▷ $\gamma = 0.333333\ (m = 1,\ p = 3)$

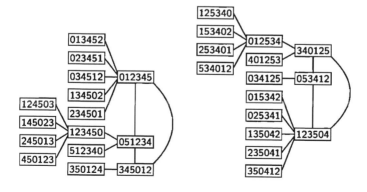

▸ $n = 6$, $e = 9$, $\delta \in \{2, \ldots, 5\}$

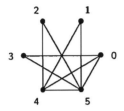

▷ $\gamma = 0.2$ ($m = 1$, $p = 5$)

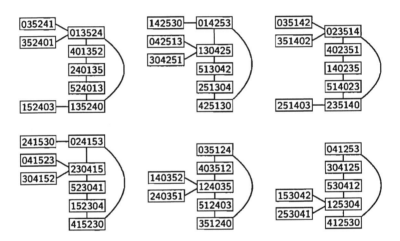

▷ $\gamma = 0.25$ ($m = 1$, $p = 4$)

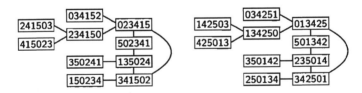

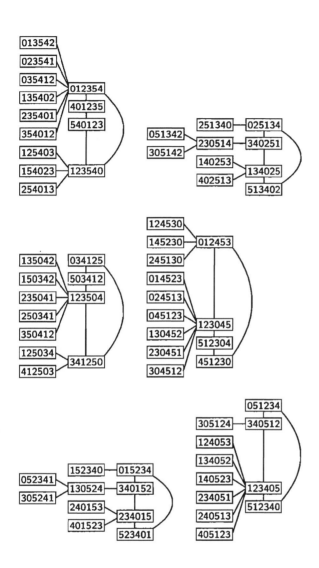

▷ $\gamma = 0.333333$ $(m = 1, p = 3)$

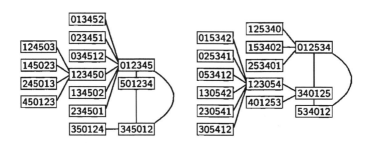

▶ $n = 6$, $e = 8$, $\delta \in \{1, \ldots, 4\}$

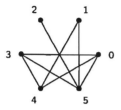

▷ $\gamma = 0.2$ $(m = 1, p = 5)$

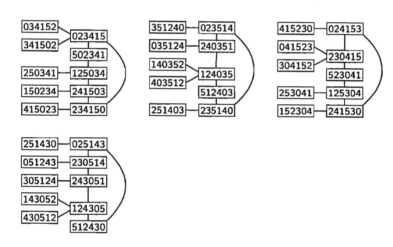

▷ $\gamma = 0.25$ $(m = 1, p = 4)$

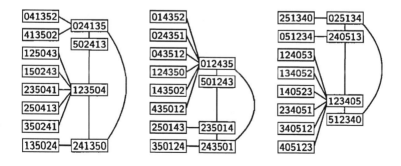

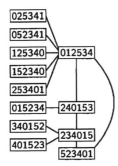

▷ $\gamma = 0.333333\ (m = 1,\ p = 3)$

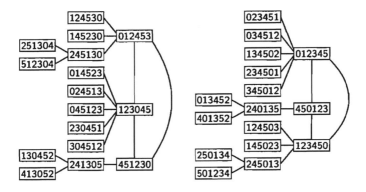

▶ $n = 6$, $e = 8$, $\delta \in \{1, \ldots, 4\}$

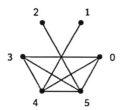

▷ $\gamma = 0.25$ $(m = 1, p = 4)$

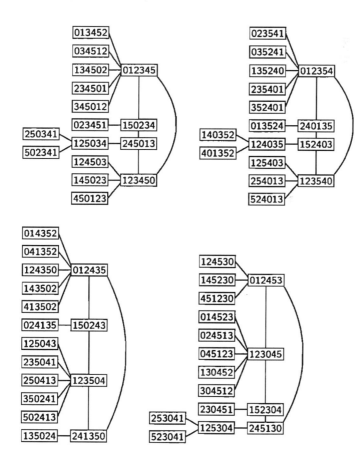

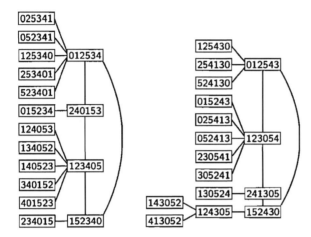

▶ $n = 6$, $e = 9$, $\delta \in \{1, \ldots, 5\}$

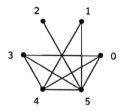

▷ $\gamma = 0.2$ $(m = 1, p = 5)$

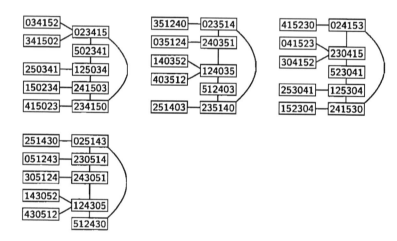

▷ $\gamma = 0.25$ $(m = 1, p = 4)$

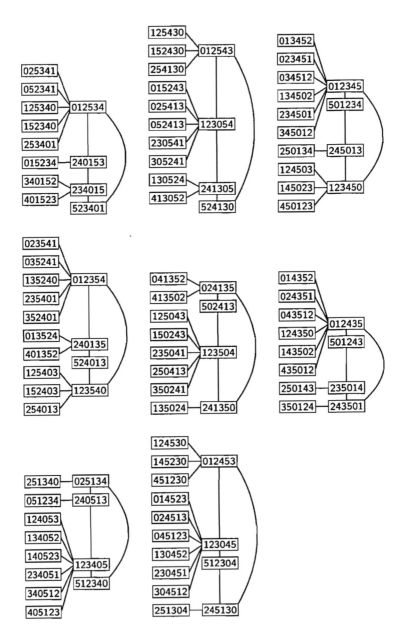

▶ $n = 6,\ e = 9,\ \delta \in \{2, \dots, 4\}$

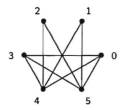

▷ $\gamma = 0.2\ (m = 1,\ p = 5)$

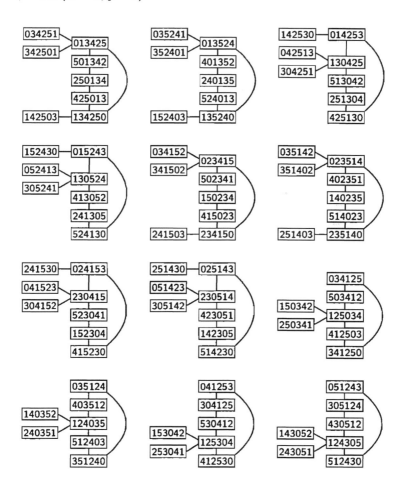

▷ $\gamma = 0.25$ $(m = 1, p = 4)$

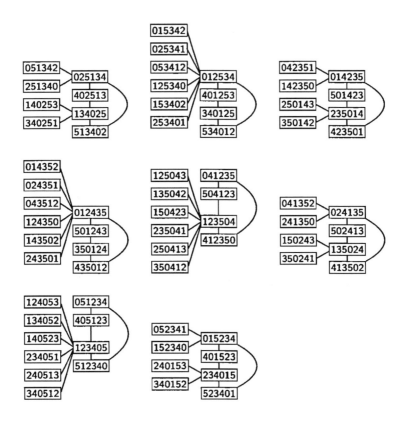

▷ $\gamma = 0.333333 \ (m = 1, \ p = 3)$

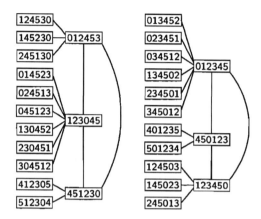

▶ $n = 6$, $e = 10$, $\delta \in \{2, \ldots, 5\}$

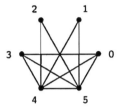

▷ $\gamma = 0.2$ $(m = 1, p = 5)$

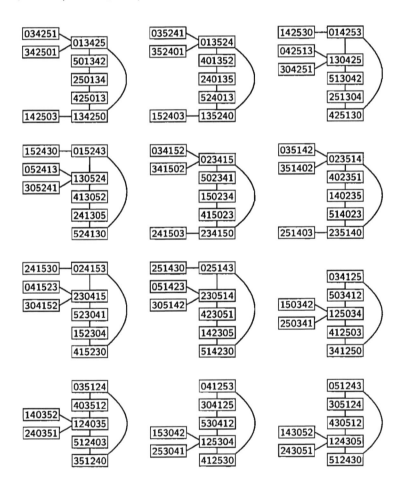

▷ $\gamma = 0.25$ $(m = 1, p = 4)$

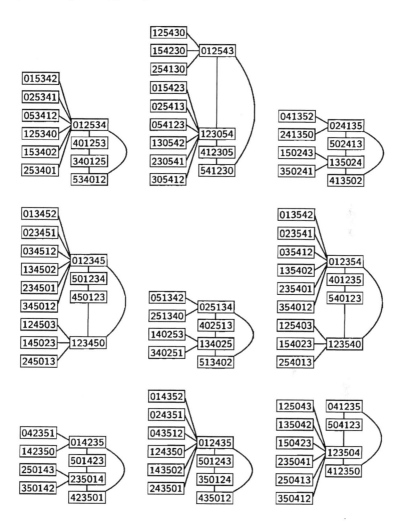

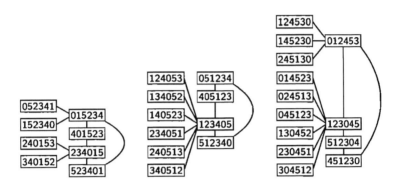

▶ $n = 6$, $e = 7$, $\delta \in \{1, \ldots, 3\}$

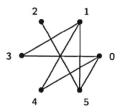

▷ $\gamma = 0.25$ $(m = 1, p = 4)$

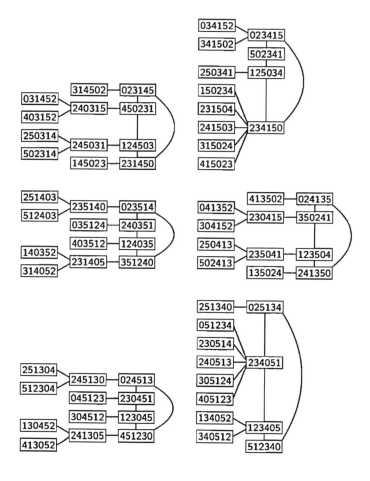

▷ $\gamma = 0.5$ $(m = 1,\ p = 2)$

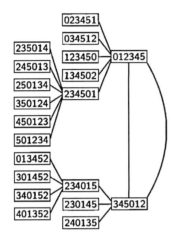

▶ $n = 6$, $e = 8$, $\delta \in \{1, \ldots, 4\}$

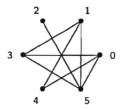

▷ $\gamma = 0.2$ $(m = 1, p = 5)$

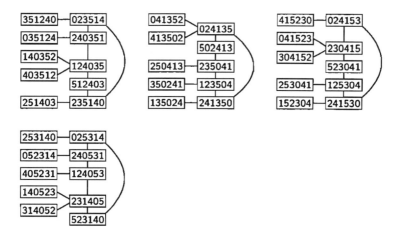

▷ $\gamma = 0.25$ $(m = 1, p = 4)$

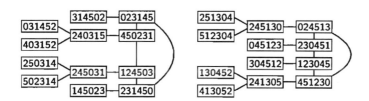

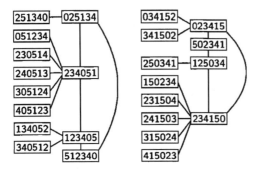

▷ $\gamma = 0.333333$ $(m = 1, p = 3)$

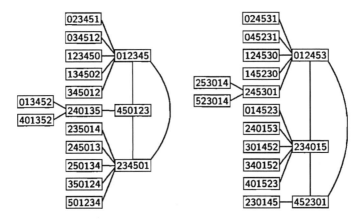

▶ $n = 6$, $e = 9$, $\delta \in \{1, \ldots, 5\}$

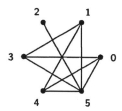

▷ $\gamma = 0.2$ $(m = 1, p = 5)$

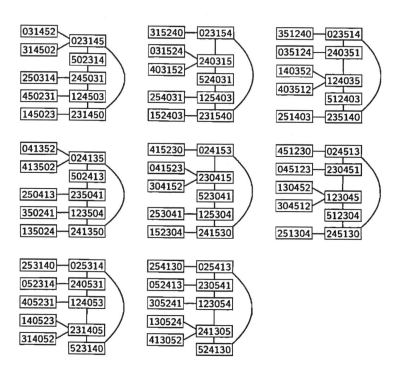

▷ $\gamma = 0.25\ (m = 1,\ p = 4)$

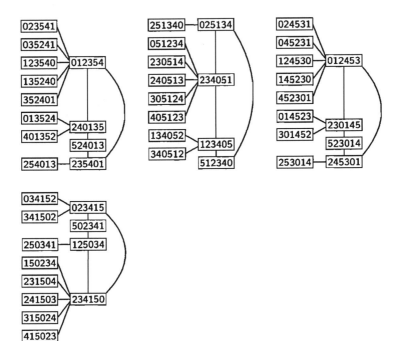

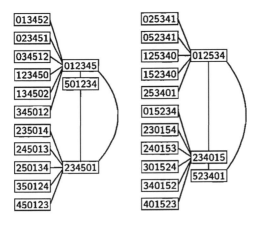

▷ $\gamma = 0.333333\ (m = 1,\ p = 3)$

▶ $n = 6,\ e = 6,\ \delta = 2$

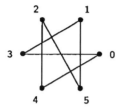

▷ $\gamma = 0.166667\ (m = 1,\ p = 6)$

031524	042513
403152	304251
240315	130425
524031	513042
152403	251304
315240	425130

▷ $\gamma = 0.333333\ (m = 2,\ p = 6)$

034251 — 013425	135240 — 013524
301425 — 350142	301524 — 340152
135042 — 123504	034152 — 023415
512304 — 451230	402315 — 450231
245130 — 024513	245031 — 124503
402513 — 340251	512403 — 351240

▷ $\gamma = 0.333333\ (m = 1,\ p = 3)$

035124	045123
145023	235014
234015	134025

▷ $\gamma = 0.5 \ (m = 1, \ p = 2)$

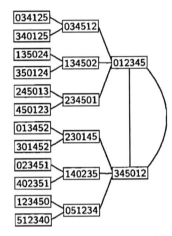

▶ $n = 6$, $e = 7$, $\delta \in \{2, 3\}$

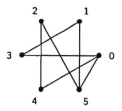

▷ $\gamma = 0.166667$ $(m = 1, p = 6)$

031524	042513
403152	304251
240315	130425
524031	513042
152403	251304
315240	425130

▷ $\gamma = 0.25$ $(m = 1, p = 4)$

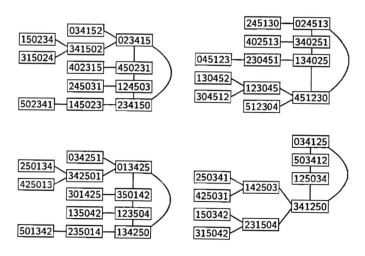

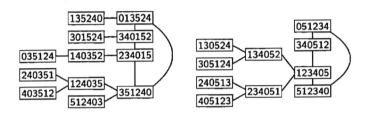

$\triangleright \quad \gamma = 0.5 \ (m = 1, \ p = 2)$

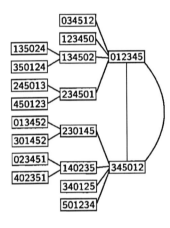

▶ $n = 6,\ e = 7,\ \delta \in \{1, \ldots, 3\}$

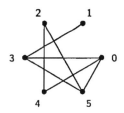

▷ $\gamma = 0.2\ (m = 1,\ p = 5)$

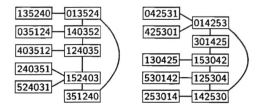

▷ $\gamma = 0.25\ (m = 1,\ p = 4)$

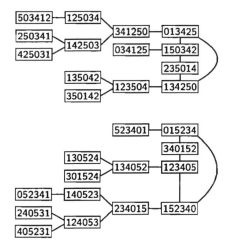

▷ $\gamma = 0.333333\ (m = 1,\ p = 3)$

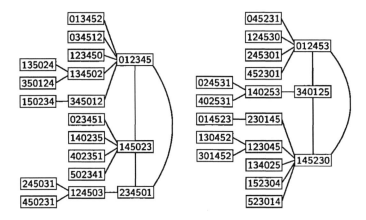

▶ $n = 6,\ e = 7,\ \delta \in \{2, 3\}$

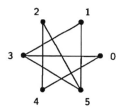

▷ $\gamma = 0.166667\ (m = 1,\ p = 6)$

031524	042513
403152	304251
240315	130425
524031	513042
152403	251304
315240	425130

▷ $\gamma = 0.2\ (m = 1,\ p = 5)$

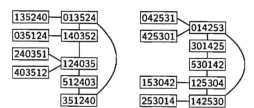

▷ $\gamma = 0.333333 \ (m = 1, \ p = 3)$

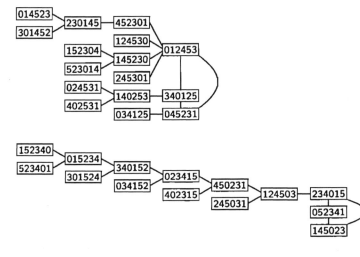

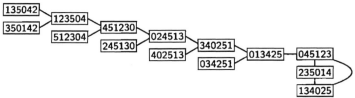

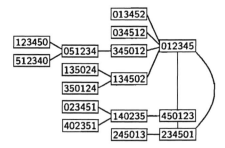

▶ $n = 6$, $e = 7$, $\delta \in \{1, \ldots, 3\}$

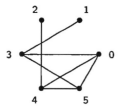

▷ $\gamma = 0.25$ $(m = 1,\ p = 4)$

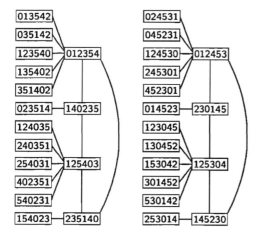

▷ $\gamma = 0.333333 \; (m = 1, \, p = 3)$

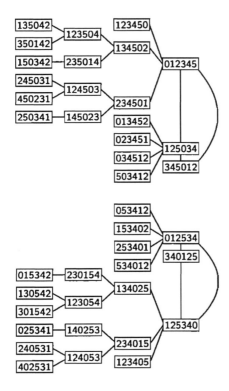

▶ $n = 6$, $e = 7$, $\delta \in \{1, \ldots, 3\}$

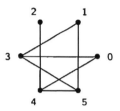

▷ $\gamma = 0.2$ ($m = 1$, $p = 5$)

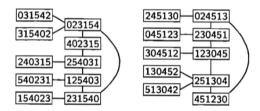

▷ $\gamma = 0.25$ ($m = 1$, $p = 4$)

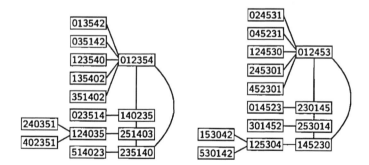

▷ $\gamma = 0.333333 \ (m = 1, \ p = 3)$

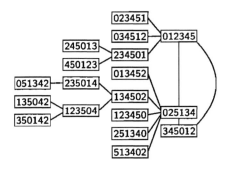

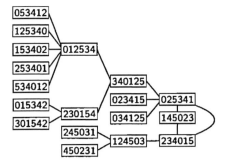

▶ $n = 6, e = 8, \delta \in \{2, \ldots, 4\}$

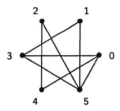

▷ $\gamma = 0.166667\ (m = 1,\ p = 6)$

031524	042513
403152	304251
240315	130425
524031	513042
152403	251304
315240	425130

▷ $\gamma = 0.2\ (m = 1,\ p = 5)$

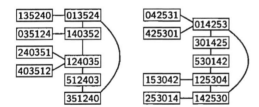

▷ $\gamma = 0.25$ $(m = 1,\ p = 4)$

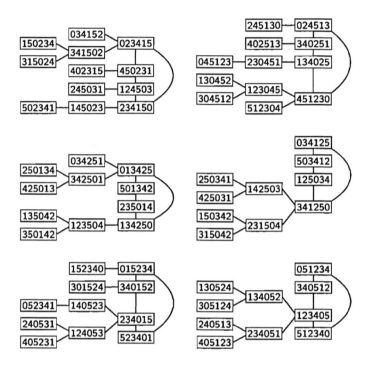

▷ $\gamma = 0.333333$ $(m = 1,\ p = 3)$

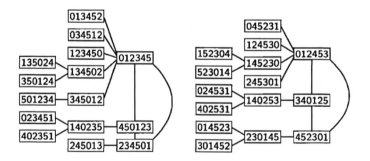

▶ $n = 6,\ e = 8,\ \delta \in \{1, \ldots, 4\}$

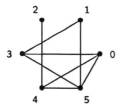

▷ $\gamma = 0.2\ (m = 1,\ p = 5)$

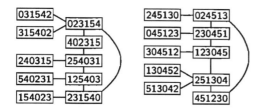

▷ $\gamma = 0.25\ (m = 1,\ p = 4)$

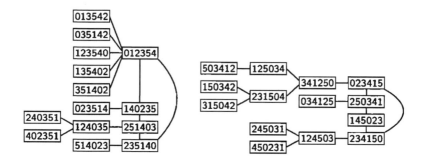

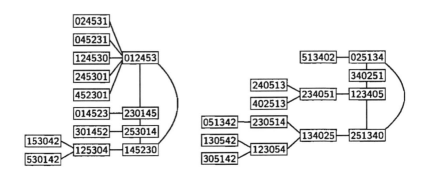

▷ $\gamma = 0.333333 \ (m = 1, \ p = 3)$

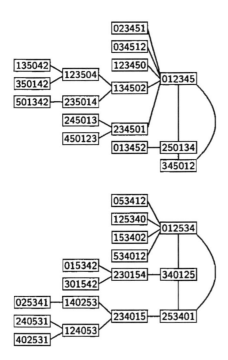

▸ $n = 6$, $e = 8$, $\delta \in \{2, \ldots, 4\}$

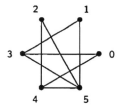

▷ $\gamma = 0.166667$ $(m = 1,\, p = 6)$

```
031524     042513
403152     304251
240315     130425
524031     513042
152403     251304
315240     425130
```

▷ $\gamma = 0.2$ $(m = 1,\, p = 5)$

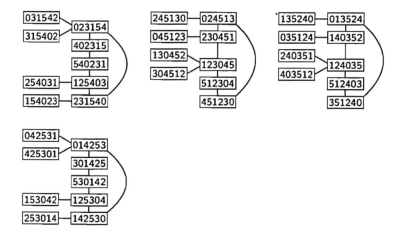

▷ $\gamma = 0.25$ $(m = 1, p = 4)$

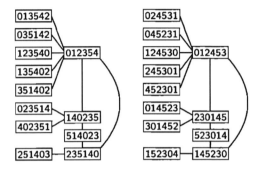

▷ $\gamma = 0.333333$ $(m = 1, p = 3)$

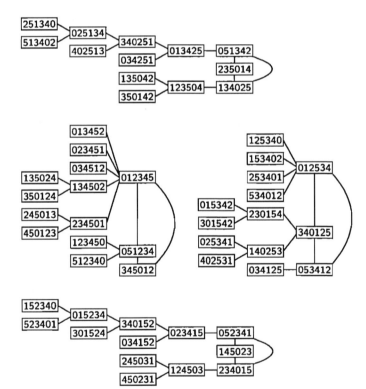

▶ $n = 6, e = 9, \delta \in \{2, \ldots, 5\}$

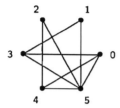

▷ $\gamma = 0.166667 \; (m = 1, p = 6)$

031524	042513
403152	304251
240315	130425
524031	513042
152403	251304
315240	425130

▷ $\gamma = 0.2 \; (m = 1, p = 5)$

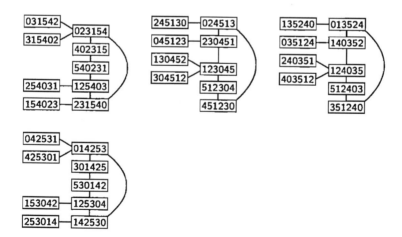

▷ $\gamma = 0.25$ $(m = 1, p = 4)$

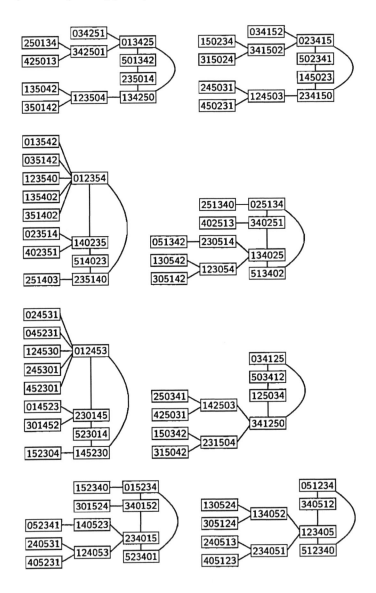

▷ $\gamma = 0.333333$ $(m = 1, p = 3)$

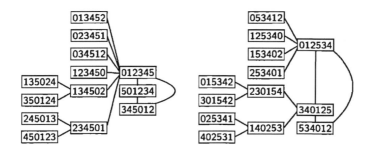

▶ $n = 6$, $e = 8$, $\delta \in \{2, 3\}$

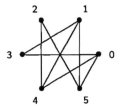

▷ $\gamma = 0.166667$ ($m = 1$, $p = 6$)

031425	031524	042513	052413
503142	403152	304251	305241
250314	240315	130425	130524
425031	524031	513042	413052
142503	152403	251304	241305
314250	315240	425130	524130

▷ $\gamma = 0.25$ ($m = 1$, $p = 4$)

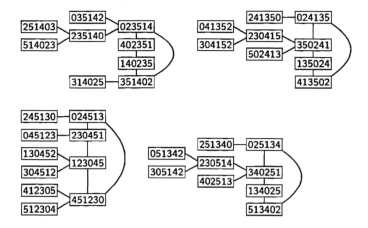

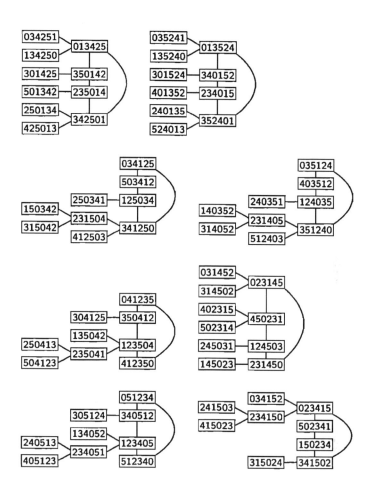

▷ $\gamma = 0.5\ (m = 1,\ p = 2)$

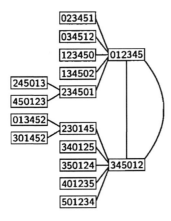

▶ $n = 6$, $e = 8$, $\delta \in \{1, \ldots, 3\}$

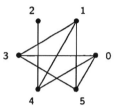

▷ $\gamma = 0.2$ $(m = 1, p = 5)$

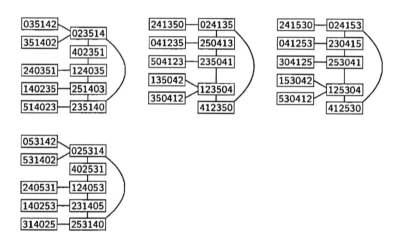

▷ $\gamma = 0.25$ $(m = 1, p = 4)$

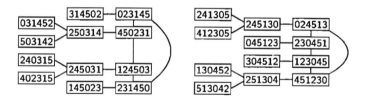

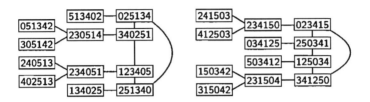

▷ $\gamma = 0.333333\ (m = 1,\ p = 3)$

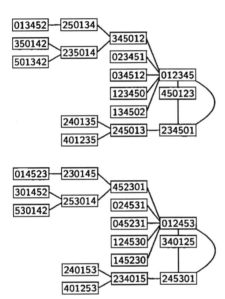

▶ $n = 6$, $e = 9$, $\delta \in \{2, \ldots, 4\}$

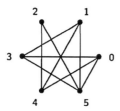

▷ $\gamma = 0.166667$ ($m = 1$, $p = 6$)

031425	031524	042513	052413
503142	403152	304251	305241
250314	240315	130425	130524
425031	524031	513042	413052
142503	152403	251304	241305
314250	315240	425130	524130

▷ $\gamma = 0.2$ ($m = 1$, $p = 5$)

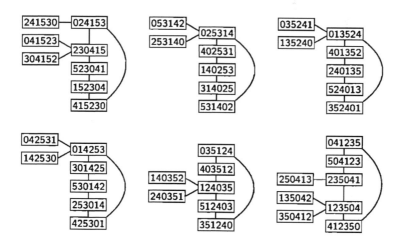

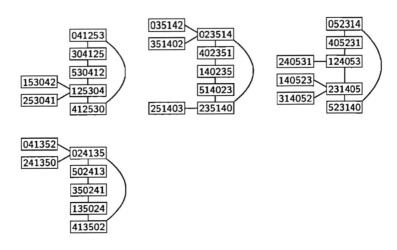

▷ $\gamma = 0.25$ $(m = 1,\ p = 4)$

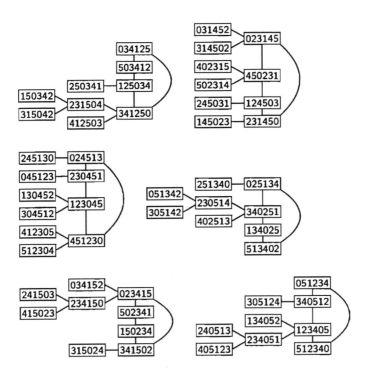

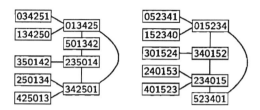

▷ $\gamma = 0.333333 \ (m = 1, \ p = 3)$

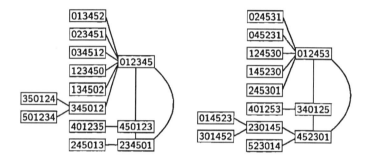

▶ $n = 6, e = 9, \delta \in \{2, \ldots, 4\}$

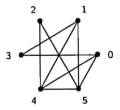

▷ $\gamma = 0.166667 \ (m = 1, p = 6)$

031425	031524	042513	052413
503142	403152	304251	305241
250314	240315	130425	130524
425031	524031	513042	413052
142503	152403	251304	241305
314250	315240	425130	524130

▷ $\gamma = 0.2 \ (m = 1, p = 5)$

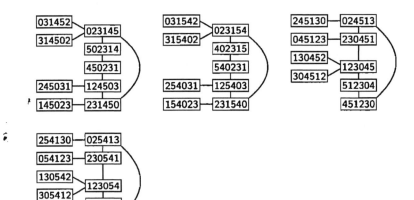

▷ $\gamma = 0.25$ $(m = 1,\, p = 4)$

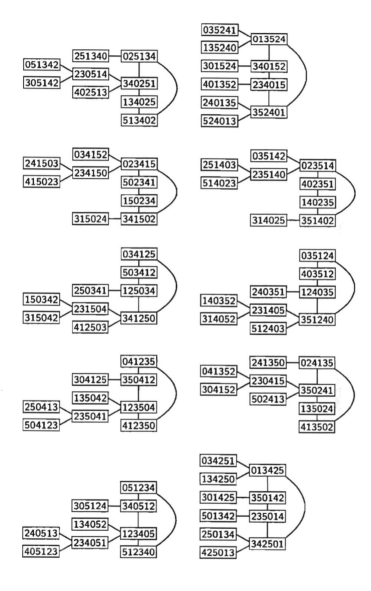

▷ $\gamma = 0.333333 \ (m = 1, \ p = 3)$

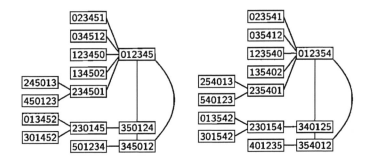

▸ $n = 6, e = 10, \delta \in \{2, \ldots, 5\}$

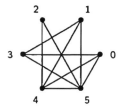

▷ $\gamma = 0.166667 \ (m = 1, p = 6)$

031425	031524	042513	052413
503142	403152	304251	305241
250314	240315	130425	130524
425031	524031	513042	413052
142503	152403	251304	241305
314250	315240	425130	524130

▷ $\gamma = 0.2 \ (m = 1, p = 5)$

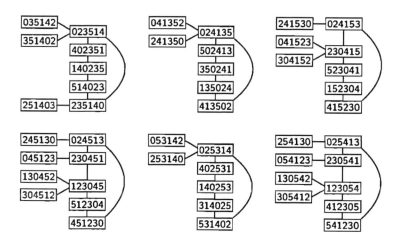

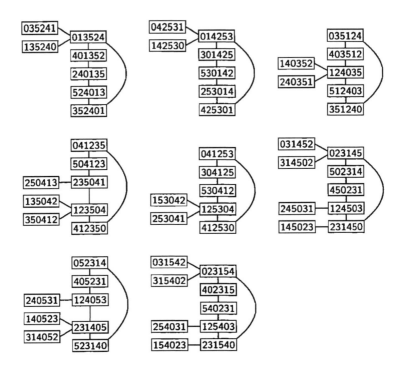

▷ $\gamma = 0.25$ $(m = 1, p = 4)$

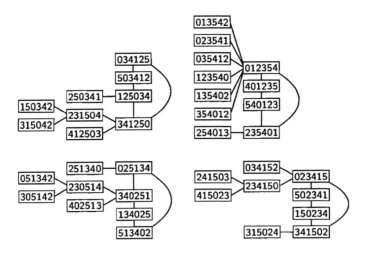

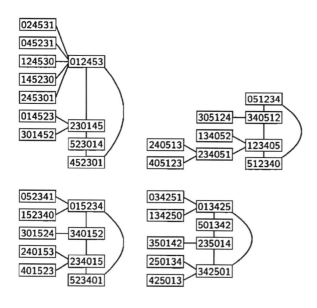

▷ $\gamma = 0.333333$ $(m = 1,\ p = 3)$

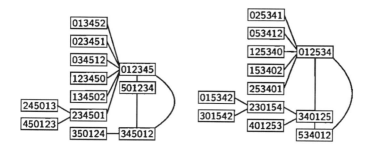

▶ $n = 6$, $e = 9$, $\delta \in \{1, \ldots, 4\}$

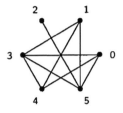

▷ $\gamma = 0.2$ $(m = 1, p = 5)$

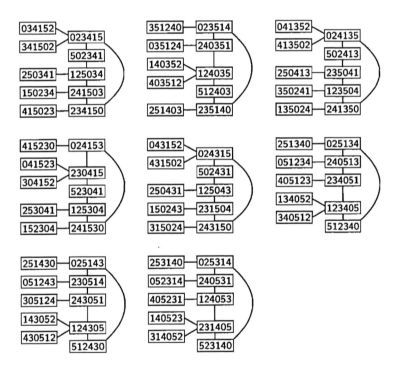

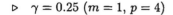

▷ $\gamma = 0.25$ $(m = 1, p = 4)$

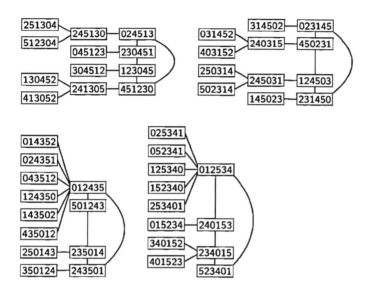

▷ $\gamma = 0.333333$ $(m = 1, p = 3)$

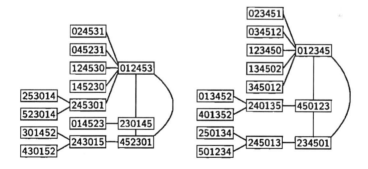

▶ $n = 6$, $e = 9$, $\delta \in \{1, \ldots, 4\}$

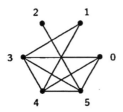

▷ $\gamma = 0.2$ $(m = 1,\ p = 5)$

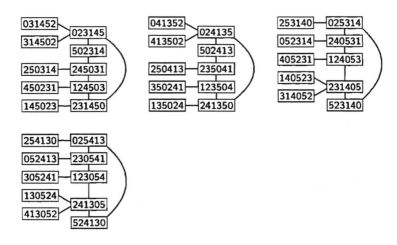

▷ $\gamma = 0.25$ $(m = 1,\ p = 4)$

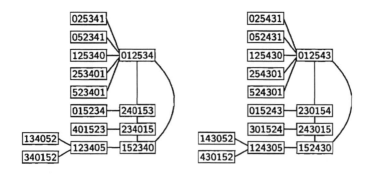

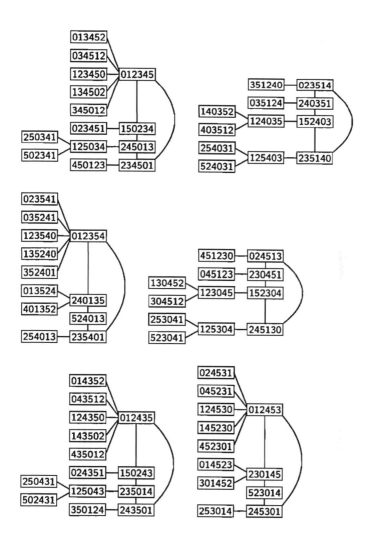

▶ $n = 6,\ e = 10,\ \delta \in \{1, \ldots, 5\}$

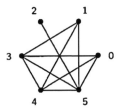

▷ $\gamma = 0.2\ (m = 1,\ p = 5)$

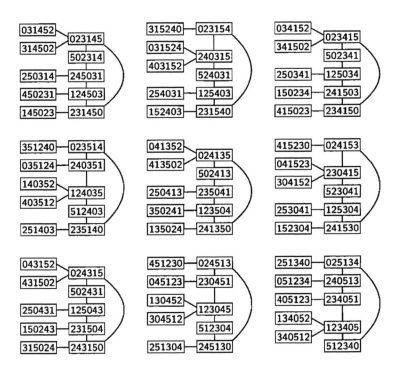

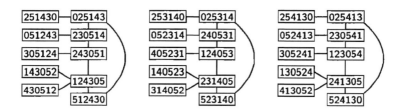

▷ $\gamma = 0.25$ $(m = 1,\ p = 4)$

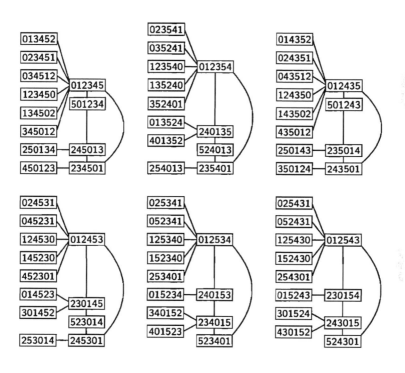

▶ $n = 6$, $e = 9$, $\delta \in \{2, \ldots, 4\}$

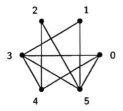

▷ $\gamma = 0.166667$ $(m = 1, p = 6)$

031524	042513
403152	304251
240315	130425
524031	513042
152403	251304
315240	425130

▷ $\gamma = 0.2$ $(m = 1, p = 5)$

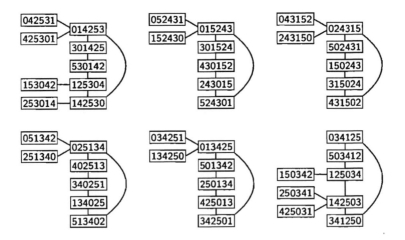

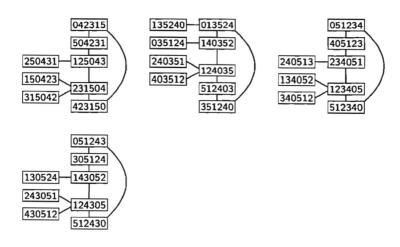

▷ $\gamma = 0.25\ (m = 1,\ p = 4)$

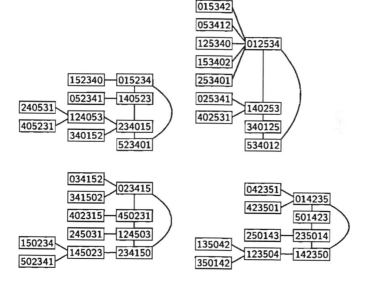

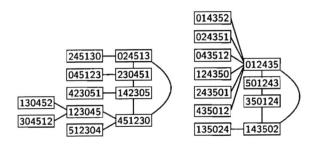

▷ $\gamma = 0.333333 \ (m = 1, \ p = 3)$

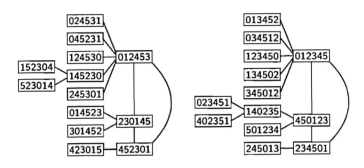

▶ $n = 6$, $e = 9$, $\delta \in \{2,\ldots,4\}$

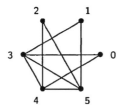

▷ $\gamma = 0.166667$ $(m = 1, p = 6)$

```
031524      042513
403152      304251
240315      130425
524031      513042
152403      251304
315240      425130
```

▷ $\gamma = 0.2$ $(m = 1, p = 5)$

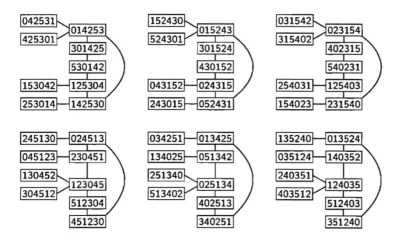

▷ $\gamma = 0.25\ (m = 1,\ p = 4)$

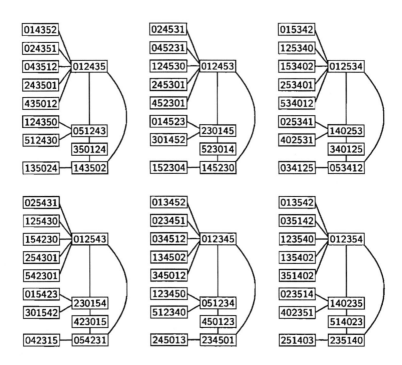

▷ $\gamma = 0.333333\ (m = 1,\ p = 3)$

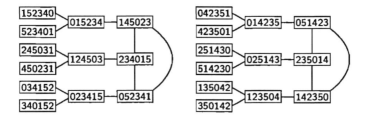

► $n = 6$, $e = 10$, $\delta \in \{2, \ldots, 5\}$

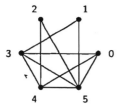

▷ $\gamma = 0.166667$ $(m = 1, p = 6)$

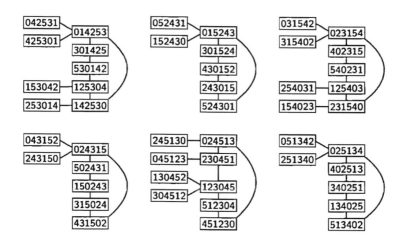

▷ $\gamma = 0.2$ $(m = 1, p = 5)$

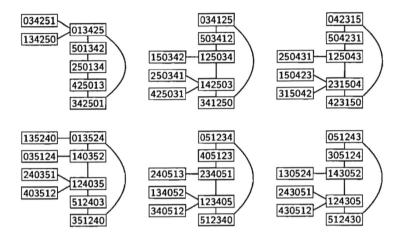

▷ $\gamma = 0.25$ $(m = 1, p = 4)$

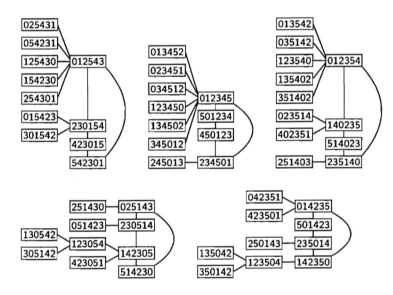

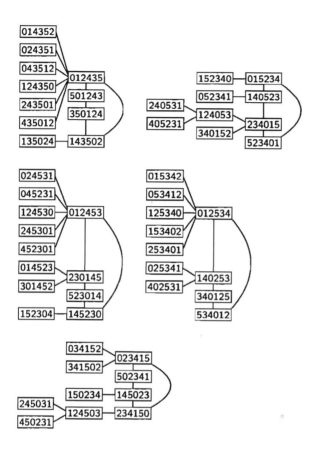

▶ $n = 6$, $e = 11$, $\delta \in \{2, \ldots, 5\}$

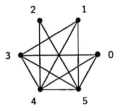

▷ $\gamma = 0.166667$ $(m = 1, p = 6)$

031425	031524	042513	052413
503142	403152	304251	305241
250314	240315	130425	130524
425031	524031	513042	413052
142503	152403	251304	241305
314250	315240	425130	524130

▷ $\gamma = 0.2$ $(m = 1, p = 5)$

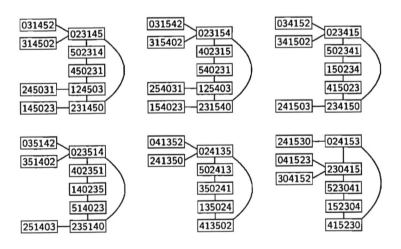

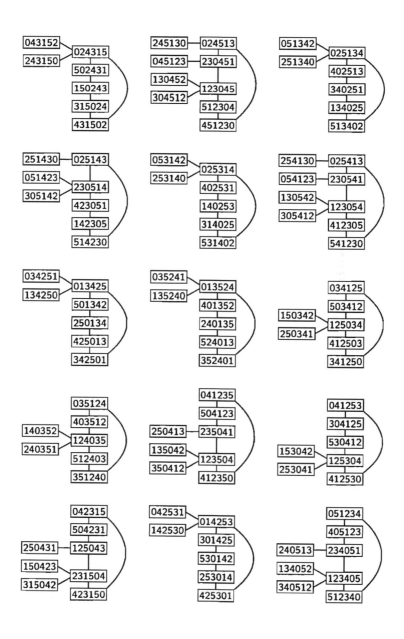

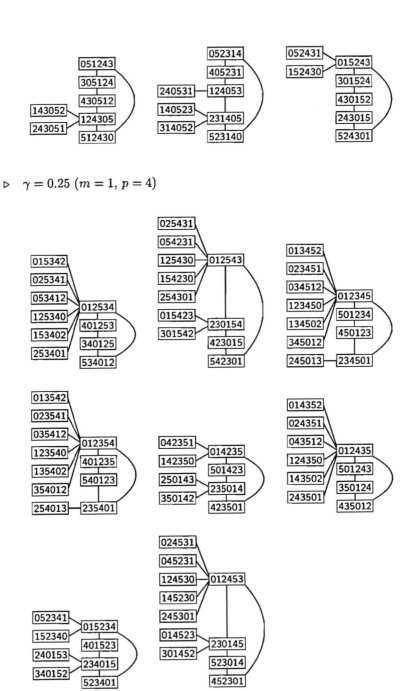

▷ $\gamma = 0.25 \; (m = 1, \, p = 4)$

▶ $n = 6,\ e = 9,\ \delta = 3$

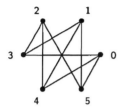

▷ $\gamma = 0.166667\ (m = 1,\ p = 6)$

031425	031524	032415	032514	041325
503142	403152	503241	403251	504132
250314	240315	150324	140325	250413
425031	524031	415032	514032	325041
142503	152403	241503	251403	132504
314250	315240	324150	325140	413250

041523	042315	042513	051324	051423
304152	504231	304251	405132	305142
230415	150423	130425	240513	230514
523041	315042	513042	324051	423051
152304	231504	251304	132405	142305
415230	423150	425130	513240	514230

052314	052413
405231	305241
140523	130524
314052	413052
231405	241305
523140	524130

▷ $\gamma = 0.25 \ (m = 1, \ p = 4)$

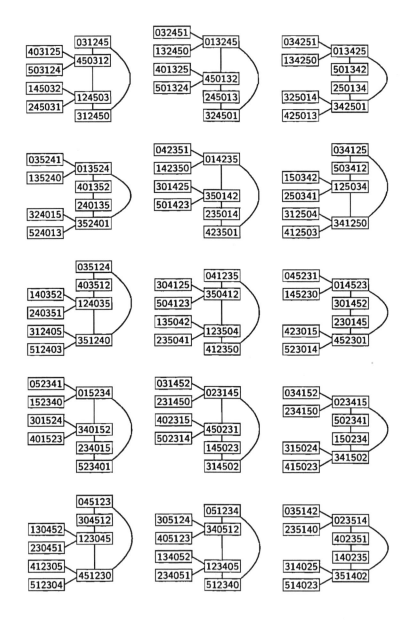

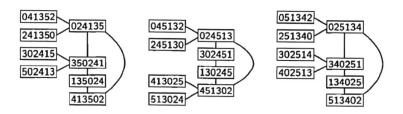

▷ $\gamma = 0.5 \ (m = 1, \ p = 2)$

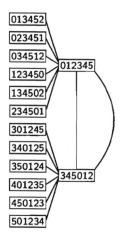

▶ $n = 6$, $e = 10$, $\delta \in \{3, 4\}$

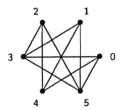

▷ $\gamma = 0.166667$ $(m = 1, p = 6)$

031425	031524	032415	032514	041325
503142	403152	503241	403251	504132
250314	240315	150324	140325	250413
425031	524031	415032	514032	325041
142503	152403	241503	251403	132504
314250	315240	324150	325140	413250

041523	042315	042513	051324	051423
304152	504231	304251	405132	305142
230415	150423	130425	240513	230514
523041	315042	513042	324051	423051
152304	231504	251304	132405	142305
415230	423150	425130	513240	514230

052314	052413
405231	305241
140523	130524
314052	413052
231405	241305
523140	524130

▷ $\gamma = 0.2\ (m = 1, p = 5)$

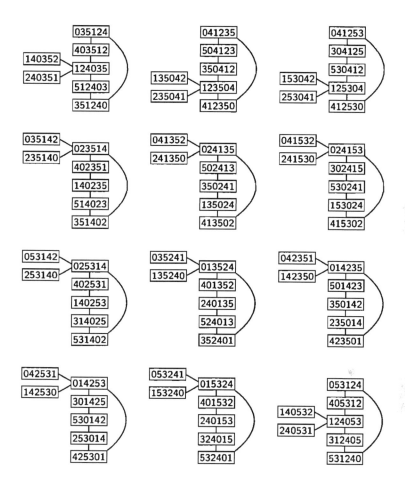

▷ $\gamma = 0.25\ (m = 1, p = 4)$

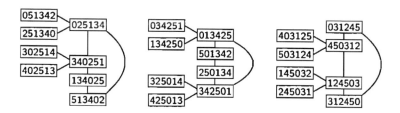

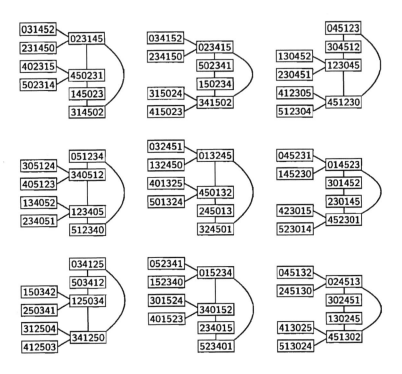

▷ $\gamma = 0.333333\ (m = 1,\ p = 3)$

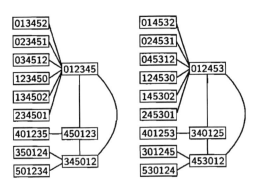

▶ $n = 6$, $e = 10$, $\delta \in \{2, \ldots, 4\}$

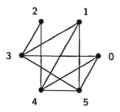

▷ $\gamma = 0.166667$ $(m = 1, p = 6)$

032415	042315	051324	051423
503241	504231	405132	305142
150324	150423	240513	230514
415032	315042	324051	423051
241503	231504	132405	142305
324150	423150	513240	514230

▷ $\gamma = 0.2$ $(m = 1, p = 5)$

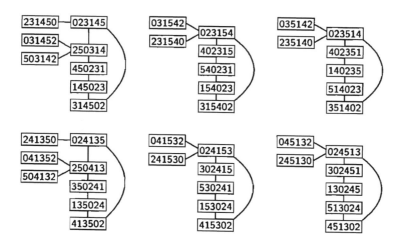

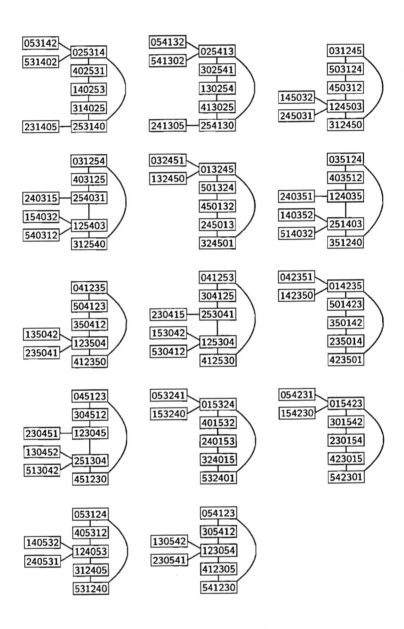

▷ $\gamma = 0.25$ $(m = 1, p = 4)$

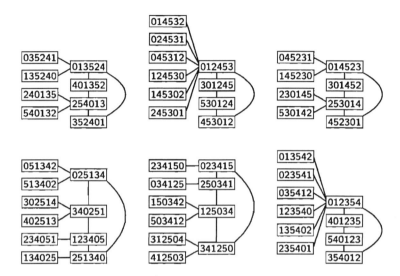

▷ $\gamma = 0.333333$ $(m = 1, p = 3)$

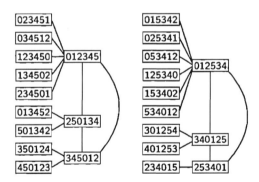

▶ $n = 6$, $e = 11$, $\delta \in \{3, \dots, 5\}$

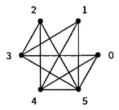

▷ $\gamma = 0.166667$ $(m = 1,\ p = 6)$

031425	031524	032415	032514	041325
503142	403152	503241	403251	504132
250314	240315	150324	140325	250413
425031	524031	415032	514032	325041
142503	152403	241503	251403	132504
314250	315240	324150	325140	413250

041523	042315	042513	051324	051423
304152	504231	304251	405132	305142
230415	150423	130425	240513	230514
523041	315042	513042	324051	423051
152304	231504	251304	132405	142305
415230	423150	425130	513240	514230

052314	052413
405231	305241
140523	130524
314052	413052
231405	241305
523140	524130

▷ $\gamma = 0.2$ $(m = 1,\ p = 5)$

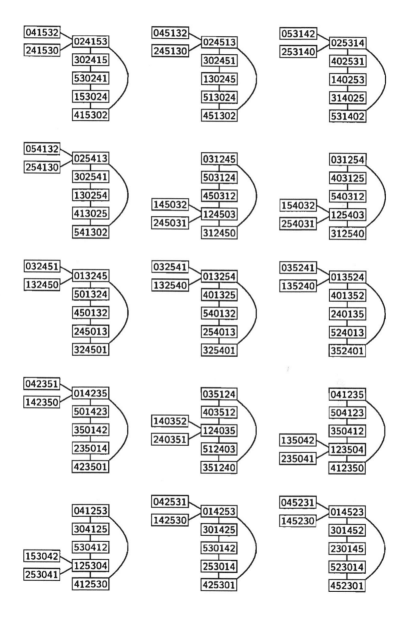

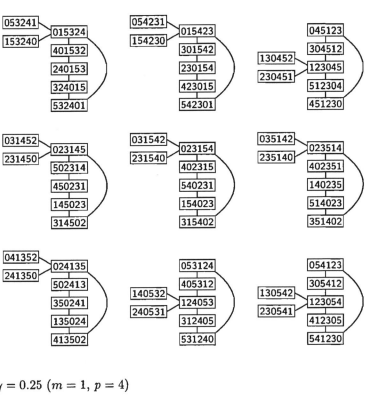

▷ $\gamma = 0.25$ $(m = 1,\ p = 4)$

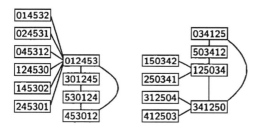

▷ $\gamma = 0.333333$ $(m = 1, p = 3)$

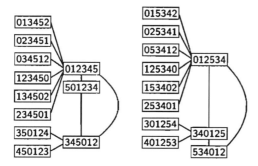

▶ $n = 6,\ e = 12,\ \delta \in \{3, \dots, 5\}$

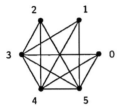

▷ $\gamma = 0.166667\ (m = 1,\ p = 6)$

031425	031524	032415	032514	041325
503142	403152	503241	403251	504132
250314	240315	150324	140325	250413
425031	524031	415032	514032	325041
142503	152403	241503	251403	132504
314250	315240	324150	325140	413250

041523	042315	042513	051324	051423
304152	504231	304251	405132	305142
230415	150423	130425	240513	230514
523041	315042	513042	324051	423051
152304	231504	251304	132405	142305
415230	423150	425130	513240	514230

052314	052413
405231	305241
140523	130524
314052	413052
231405	241305
523140	524130

▷ $\gamma = 0.2\ (m = 1,\ p = 5)$

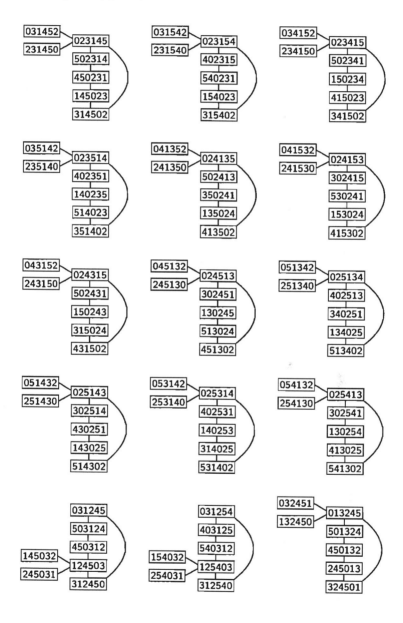

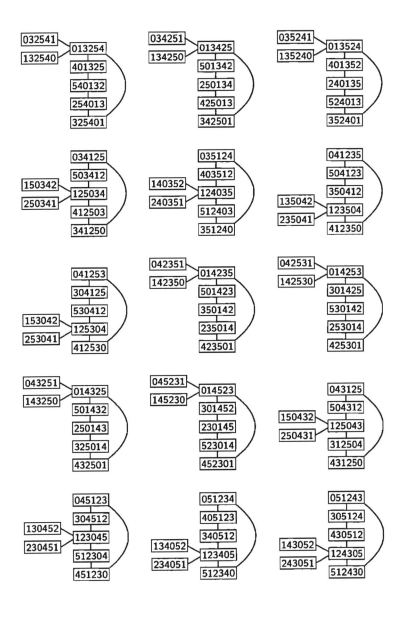

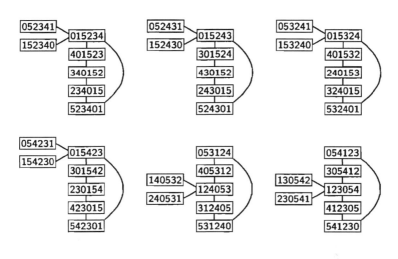

▷ $\gamma = 0.25\ (m = 1,\ p = 4)$

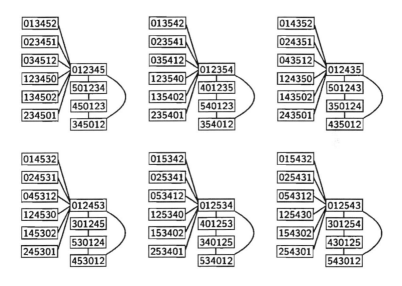

▶ $n = 6, e = 7, \delta \in \{1, \ldots, 3\}$

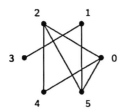

▷ $\gamma = 0.25$ $(m = 1, p = 4)$

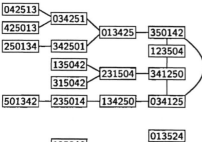

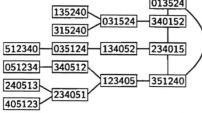

▷ $\gamma = 0.333333$ $(m = 1, p = 3)$

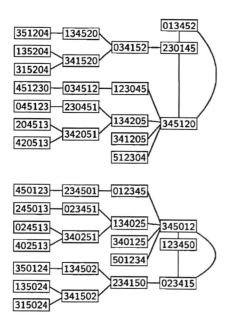

▶ $n = 6, e = 7, \delta \in \{2, 3\}$

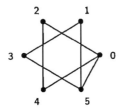

▷ $\gamma = 0.333333 \ (m = 1, p = 3)$

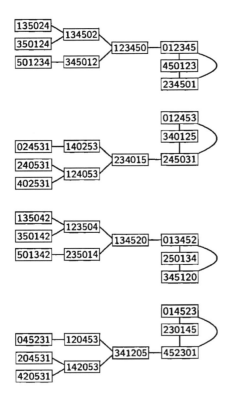

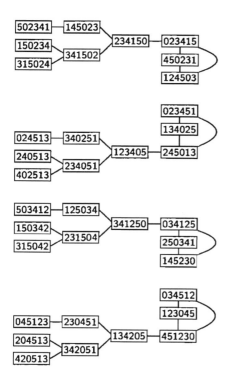

▶ $n = 6$, $e = 8$, $\delta \in \{2, \ldots, 4\}$

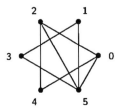

▷ $\gamma = 0.25$ $(m = 1, p = 4)$

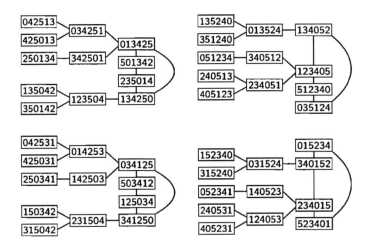

▷ $\gamma = 0.333333$ $(m = 1, p = 3)$

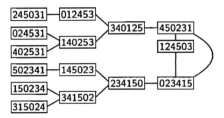

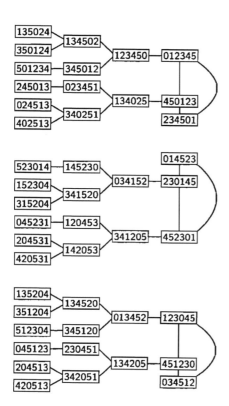

▶ $n = 6,\ e = 8,\ \delta \in \{1, \ldots, 4\}$

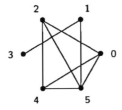

▷ $\gamma = 0.25\ (m = 1,\ p = 4)$

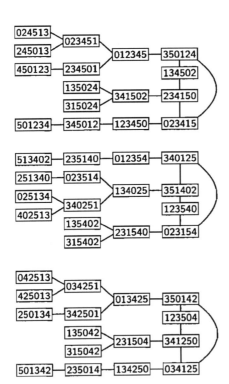

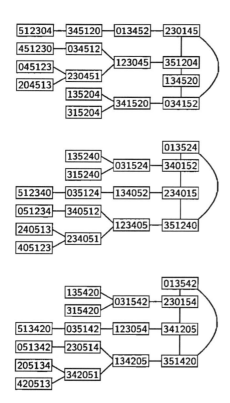

▶ $n = 6$, $e = 9$, $\delta \in \{2, \ldots, 5\}$

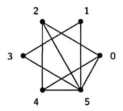

▷ $\gamma = 0.25$ $(m = 1,\ p = 4)$

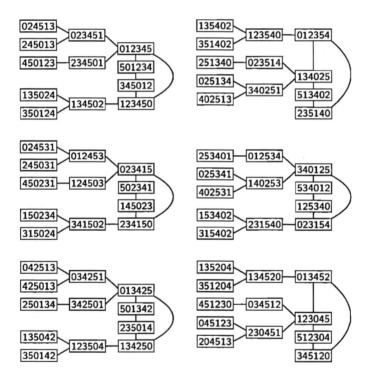

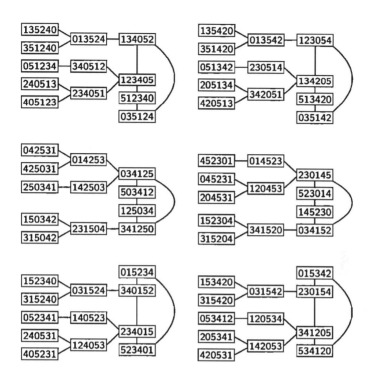

▶ $n = 6$, $e = 8$, $\delta \in \{2, 3\}$

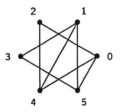

▷ $\gamma = 0.166667$ $(m = 1, p = 6)$

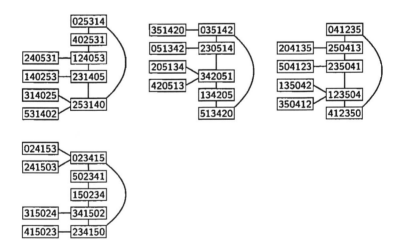

▷ $\gamma = 0.2$ $(m = 1, p = 5)$

▷ $\gamma = 0.25$ $(m = 1, p = 4)$

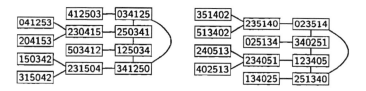

▷ $\gamma = 0.333333$ $(m = 1, p = 3)$

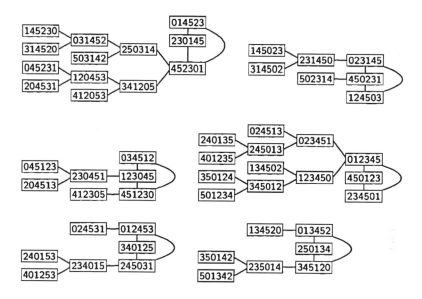

▶ $n = 6,\ e = 9,\ \delta \in \{2, \ldots, 4\}$

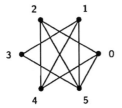

▷ $\gamma = 0.166667\ (m = 1,\ p = 6)$

024135	025314	041352	053142
502413	402531	204135	205314
350241	140253	520413	420531
135024	314025	352041	142053
413502	531402	135204	314205
241350	253140	413520	531420

▷ $\gamma = 0.2\ (m = 1,\ p = 5)$

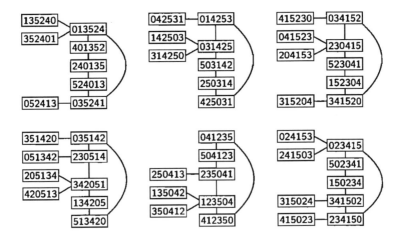

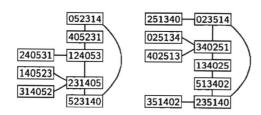

▷ $\gamma = 0.25 \ (m = 1, \ p = 4)$

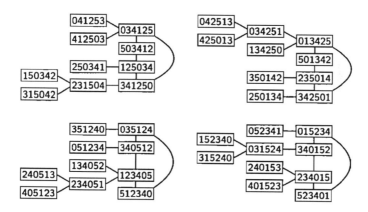

▷ $\gamma = 0.333333 \ (m = 1, \ p = 3)$

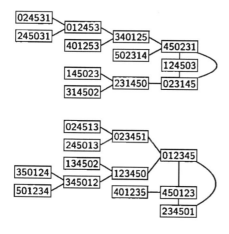

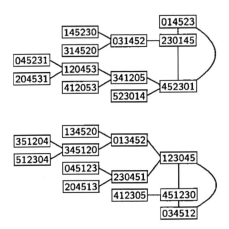

▸ $n = 6$, $e = 9$, $\delta \in \{1, \ldots, 4\}$

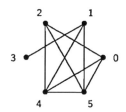

▷ $\gamma = 0.2$ $(m = 1, p = 5)$

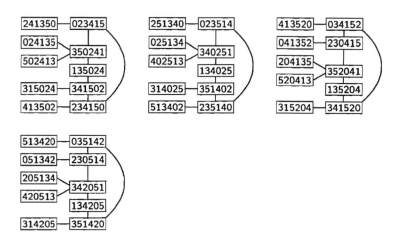

▷ $\gamma = 0.25$ $(m = 1, p = 4)$

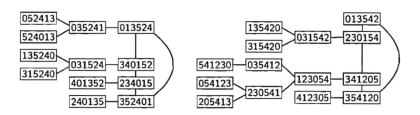

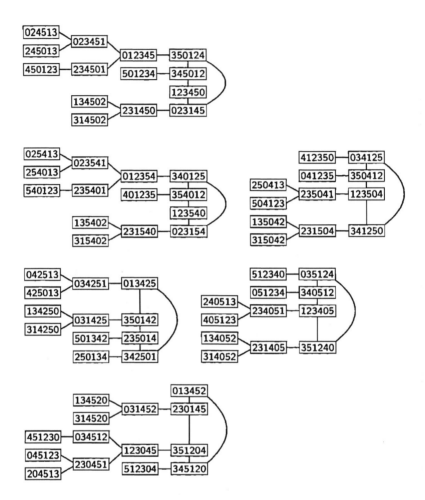

▶ $n = 6$, $e = 9$, $\delta \in \{2, \ldots, 4\}$

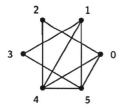

▷ $\gamma = 0.166667$ $(m = 1, p = 6)$

```
024135      053142
502413      205314
350241      420531
135024      142053
413502      314205
241350      531420
```

▷ $\gamma = 0.2$ $(m = 1, p = 5)$

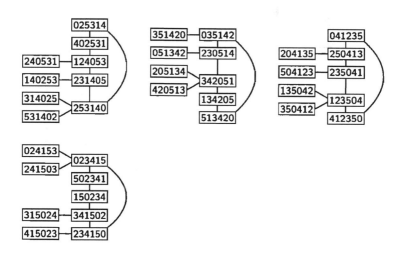

▷ $\gamma = 0.25$ $(m = 1, p = 4)$

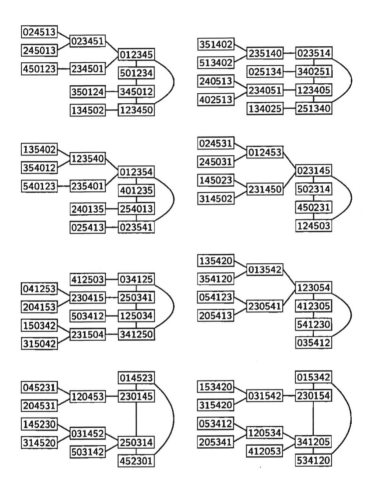

▷ $\gamma = 0.333333$ $(m = 1, p = 3)$

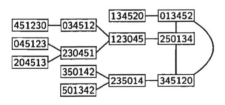

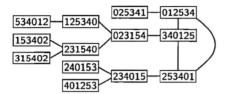

▶ $n = 6,\ e = 9,\ \delta \in \{2, \ldots, 4\}$

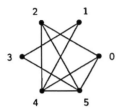

▷ $\gamma = 0.166667\ (m = 1,\ p = 6)$

024135	025314	041352	053142
502413	402531	204135	205314
350241	140253	520413	420531
135024	314025	352041	142053
413502	531402	135204	314205
241350	253140	413520	531420

▷ $\gamma = 0.2\ (m = 1,\ p = 5)$

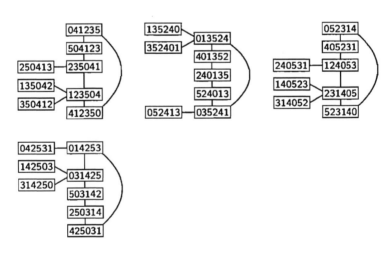

▷ $\gamma = 0.25$ $(m = 1, p = 4)$

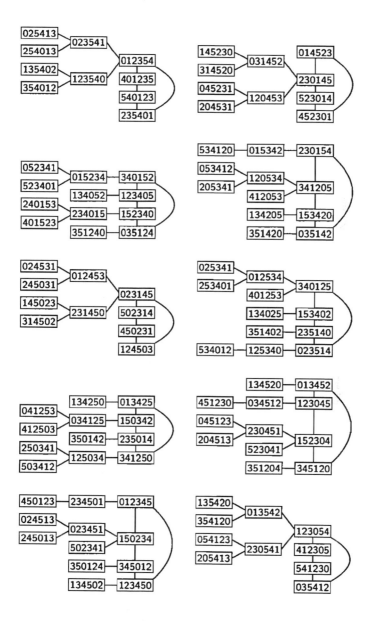

▶ $n = 6$, $e = 10$, $\delta \in \{2, \dots, 5\}$

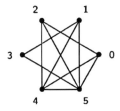

▷ $\gamma = 0.166667$ $(m = 1, p = 6)$

024135	025314	041352	053142
502413	402531	204135	205314
350241	140253	520413	420531
135024	314025	352041	142053
413502	531402	135204	314205
241350	253140	413520	531420

▷ $\gamma = 0.2$ $(m = 1, p = 5)$

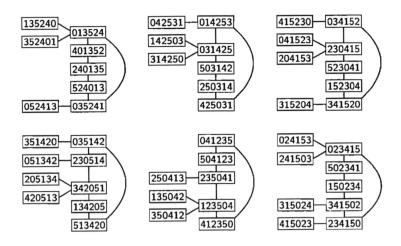

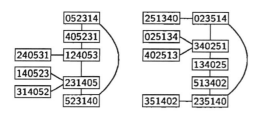

▷ $\gamma = 0.25 \ (m = 1, \ p = 4)$

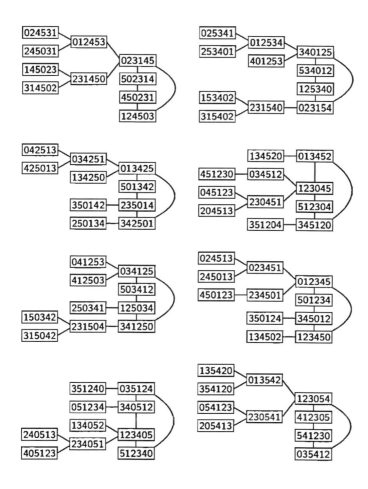

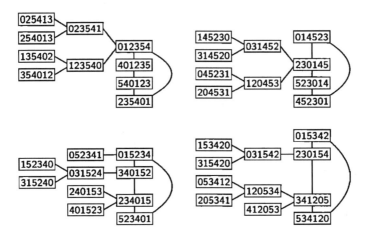

▶ $n = 6$, $e = 10$, $\delta \in \{3, 4\}$

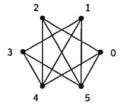

▷ $\gamma = 0.166667$ $(m = 1, p = 6)$

024135	024315	025134	025314	041352
502413	502431	402513	402531	204135
350241	150243	340251	140253	520413
135024	315024	134025	314025	352041
413502	431502	513402	531402	135204
241350	243150	251340	253140	413520

043152	051342	053142
204315	205134	205314
520431	420513	420531
152043	342051	142053
315204	134205	314205
431520	513420	531420

▷ $\gamma = 0.2$ $(m = 1, p = 5)$

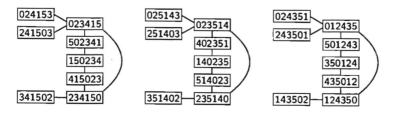

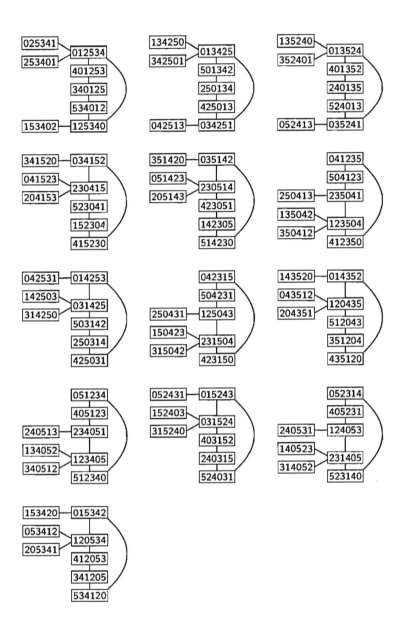

▷ $\gamma = 0.25$ $(m = 1, p = 4)$

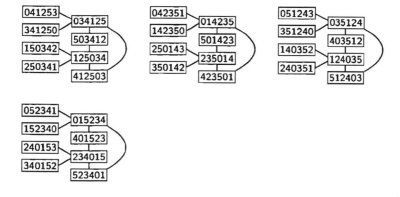

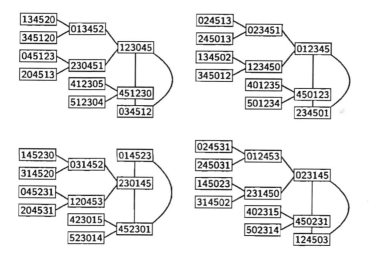

▷ $\gamma = 0.333333$ $(m = 1, p = 3)$

▶ $n = 6,\ e = 11,\ \delta \in \{3, \ldots, 5\}$

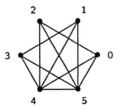

▷ $\gamma = 0.166667\ (m = 1,\ p = 6)$

024135	024315	025134	025314	041352
502413	502431	402513	402531	204135
350241	150243	340251	140253	520413
135024	315024	134025	314025	352041
413502	431502	513402	531402	135204
241350	243150	251340	253140	413520

043152	051342	053142
204315	205134	205314
520431	420513	420531
152043	342051	142053
315204	134205	314205
431520	513420	531420

▷ $\gamma = 0.2\ (m = 1,\ p = 5)$

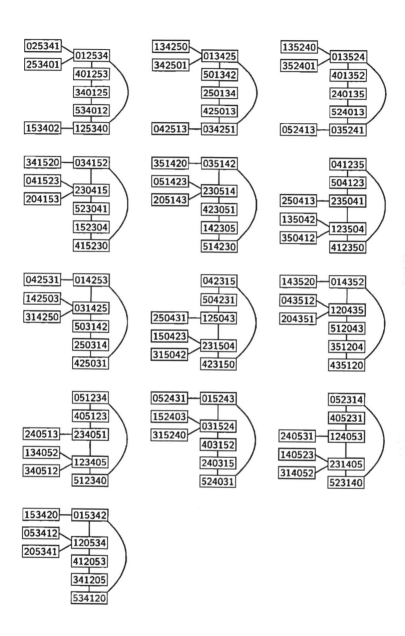

▷ $\gamma = 0.25$ $(m = 1, p = 4)$

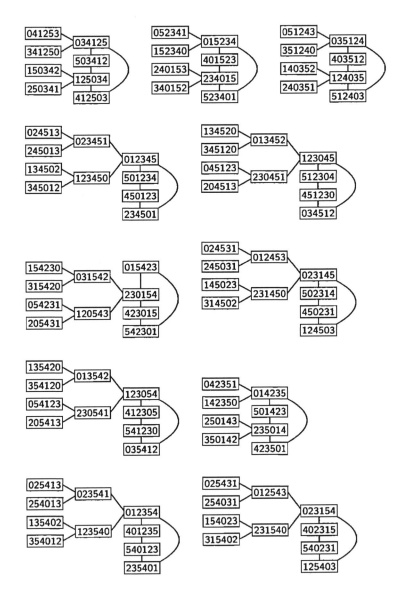

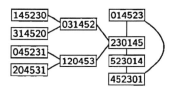

▶ $n = 6$, $e = 8$, $\delta \in \{2,3\}$

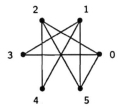

▷ $\gamma = 0.166667$ $(m = 1,\ p = 6)$

031425	052413
503142	305241
250314	130524
425031	413052
142503	241305
314250	524130

▷ $\gamma = 0.2$ $(m = 1,\ p = 5)$

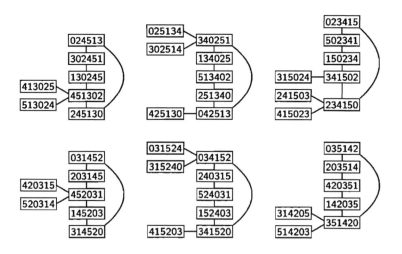

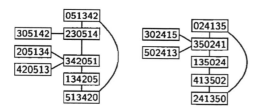

▷ $\gamma = 0.25$ $(m = 1, p = 4)$

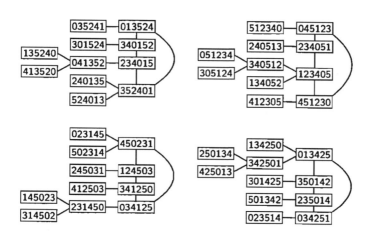

▷ $\gamma = 0.333333$ $(m = 1, p = 3)$

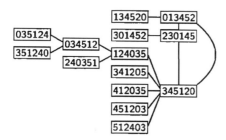

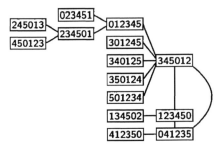

▸ $n = 6$, $e = 8$, $\delta \in \{2,3\}$

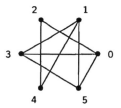

▷ $\gamma = 0.166667$ $(m = 1, p = 6)$

024135	024153	035142	053142
502413	302415	203514	205314
350241	530241	420351	420531
135024	153024	142035	142053
413502	415302	514203	314205
241350	241530	351420	531420

▷ $\gamma = 0.2$ $(m = 1, p = 5)$

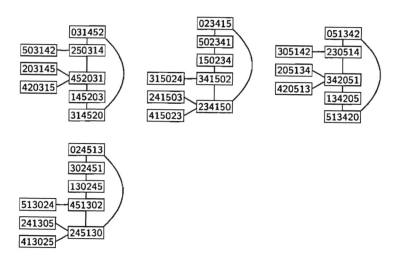

▷ $\gamma = 0.25$ $(m = 1,\ p = 4)$

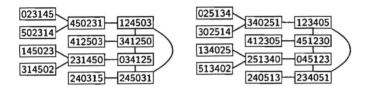

▷ $\gamma = 0.333333$ $(m = 1,\ p = 3)$

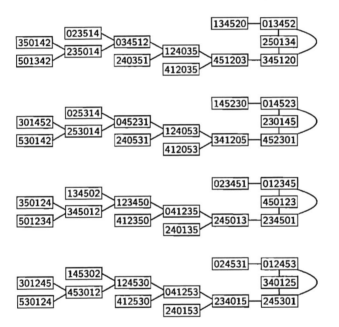

▶ $n = 6,\ e = 9,\ \delta \in \{2, \ldots, 4\}$

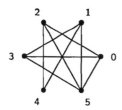

▷ $\gamma = 0.166667\ (m = 1,\ p = 6)$

024135	024153	031425	035142	052413
502413	302415	503142	203514	305241
350241	530241	250314	420351	130524
135024	153024	425031	142035	413052
413502	415302	142503	514203	241305
241350	241530	314250	351420	524130

053142
205314
420531
142053
314205
531420

▷ $\gamma = 0.2\ (m = 1,\ p = 5)$

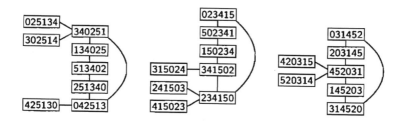

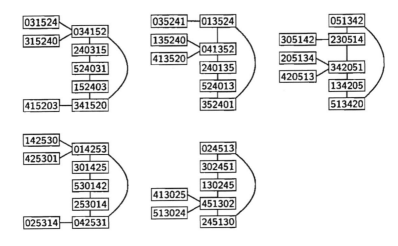

▷ $\gamma = 0.25$ ($m = 1$, $p = 4$)

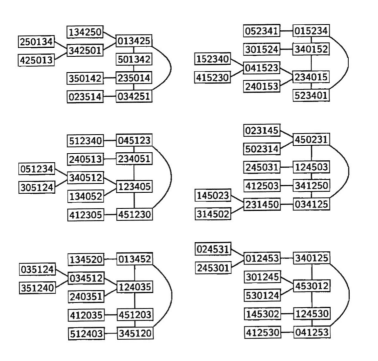

▷ $\gamma = 0.333333$ $(m = 1, p = 3)$

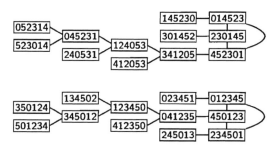

▶ $n = 6, e = 10, \delta \in \{3, \ldots, 5\}$

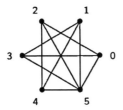

▷ $\gamma = 0.166667 \ (m = 1, p = 6)$

024135	024153	024513	025413	031425
502413	302415	302451	302541	503142
350241	530241	130245	130254	250314
135024	153024	513024	413025	425031
413502	415302	451302	541302	142503
241350	241530	245130	254130	314250

031452	031542	035142	052413	053142
203145	203154	203514	305241	205314
520314	420315	420351	130524	420531
452031	542031	142035	413052	142053
145203	154203	514203	241305	314205
314520	315420	351420	524130	531420

▷ $\gamma = 0.2 \ (m = 1, p = 5)$

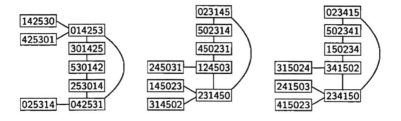

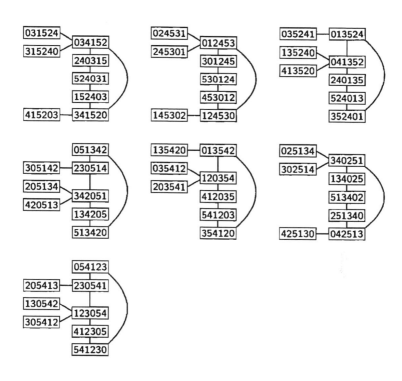

▷ $\gamma = 0.25$ $(m = 1, p = 4)$

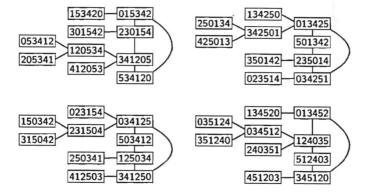

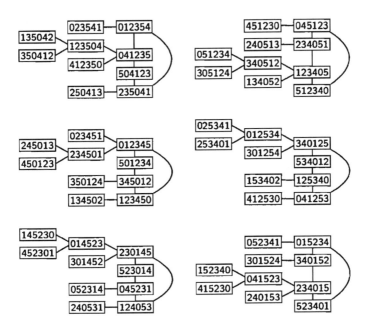

▶ $n = 6, e = 9, \delta = 3$

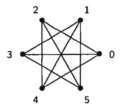

▷ $\gamma = 0.166667 \ (m = 1, \ p = 6)$

024153	025314	031524	035142	041352
302415	402531	403152	203514	204135
530241	140253	240315	420351	520413
153024	314025	524031	142035	352041
415302	531402	152403	514203	135204
241530	253140	315240	351420	413520

042513
304251
130425
513042
251304
425130

▷ $\gamma = 0.2 \ (m = 1, \ p = 5)$

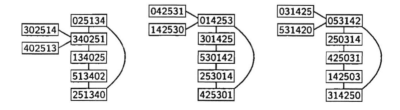

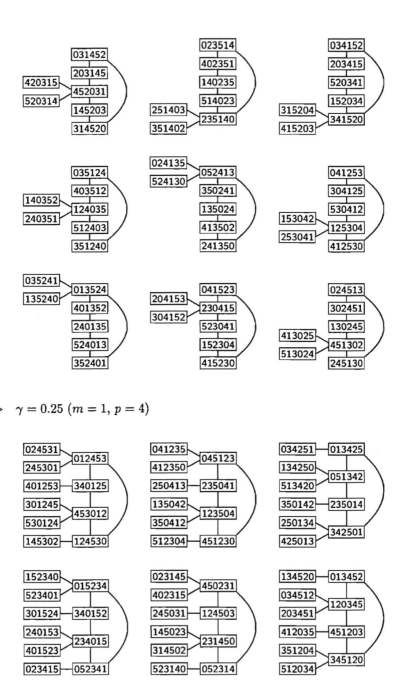

▷ $\gamma = 0.25 \; (m = 1, \, p = 4)$

▷ $\gamma = 0.333333$ $(m = 1, p = 3)$

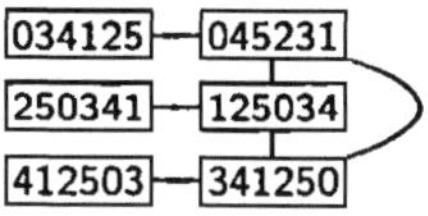

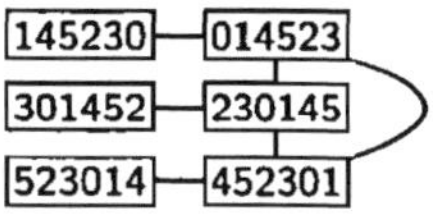

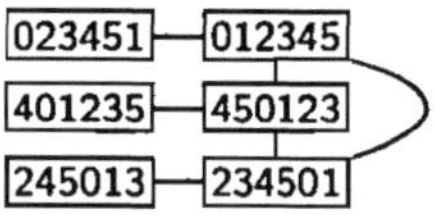

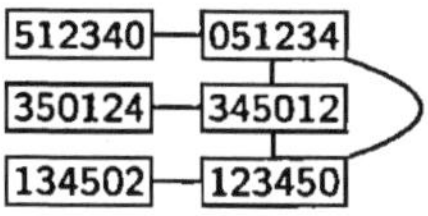

▶ $n = 6,\ e = 10,\ \delta \in \{3, 4\}$

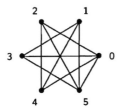

▷ $\gamma = 0.166667\ (m = 1,\ p = 6)$

024135	024153	025314	031425	031524
502413	302415	402531	503142	403152
350241	530241	140253	250314	240315
135024	153024	314025	425031	524031
413502	415302	531402	142503	152403
241350	241530	253140	314250	315240

035142	041352	042513	052413	053142
203514	204135	304251	305241	205314
420351	520413	130425	130524	420531
142035	352041	513042	413052	142053
514203	135204	251304	241305	314205
351420	413520	425130	524130	531420

▷ $\gamma = 0.2\ (m = 1,\ p = 5)$

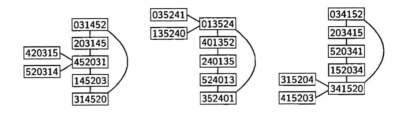

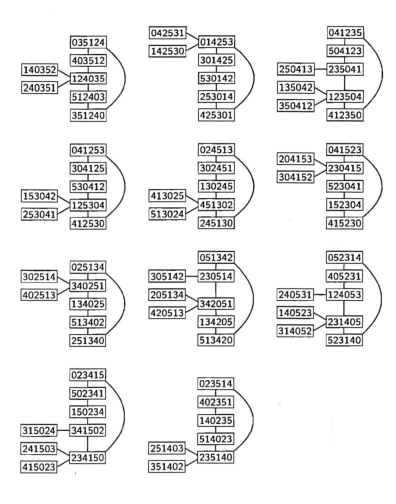

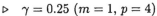

▷ $\gamma = 0.25 \ (m = 1, \ p = 4)$

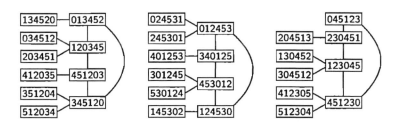

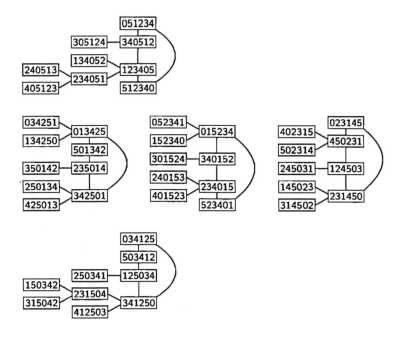

▷ $\gamma = 0.333333 \; (m = 1, \, p = 3)$

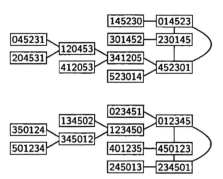

▶ $n = 6$, $e = 10$, $\delta \in \{2, \dots, 4\}$

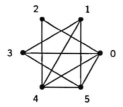

▷ $\gamma = 0.166667$ $(m = 1,\ p = 6)$

024135	024153	024513	031542	035142
502413	302415	302451	203154	203514
350241	530241	130245	420315	420351
135024	153024	513024	542031	142035
413502	415302	451302	154203	514203
241350	241530	245130	315420	351420

053142
205314
420531
142053
314205
531420

▷ $\gamma = 0.2$ $(m = 1,\ p = 5)$

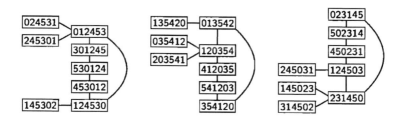

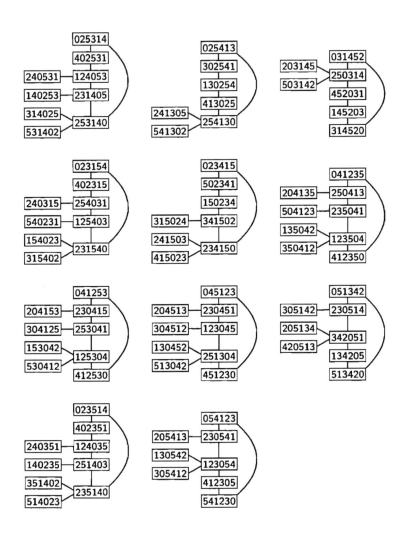

▷ $\gamma = 0.25$ $(m = 1, \, p = 4)$

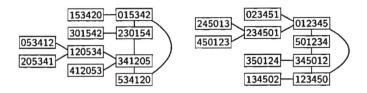

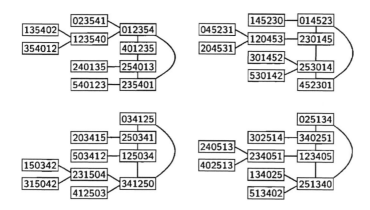

$\triangleright$ $\gamma = 0.333333$ $(m = 1,\ p = 3)$

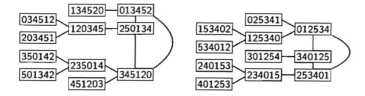

▶ $n = 6$, $e = 11$, $\delta \in \{3, \ldots, 5\}$

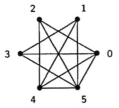

▷ $\gamma = 0.166667$ $(m = 1, p = 6)$

024135	024153	024513	025314	025413
502413	302415	302451	402531	302541
350241	530241	130245	140253	130254
135024	153024	513024	314025	413025
413502	415302	451302	531402	541302
241350	241530	245130	253140	254130

031425	031452	031524	031542	035142
503142	203145	403152	203154	203514
250314	520314	240315	420315	420351
425031	452031	524031	542031	142035
142503	145203	152403	154203	514203
314250	314520	315240	315420	351420

041352	042513	052413	053142
204135	304251	305241	205314
520413	130425	130524	420531
352041	513042	413052	142053
135204	251304	241305	314205
413520	425130	524130	531420

▷ $\gamma = 0.2 \ (m = 1, \ p = 5)$

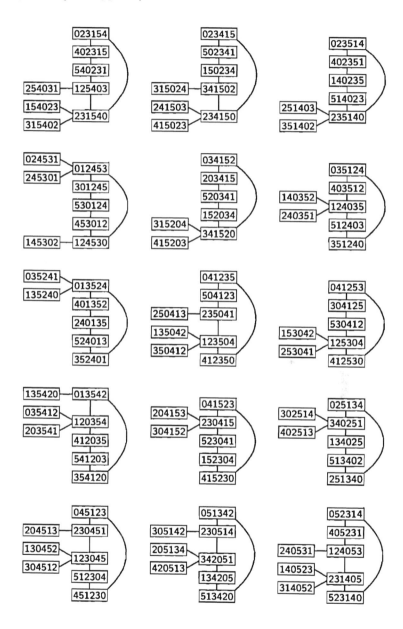

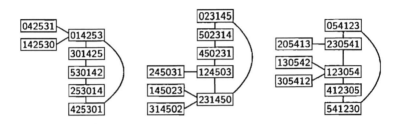

▷ $\gamma = 0.25 \ (m = 1, \ p = 4)$

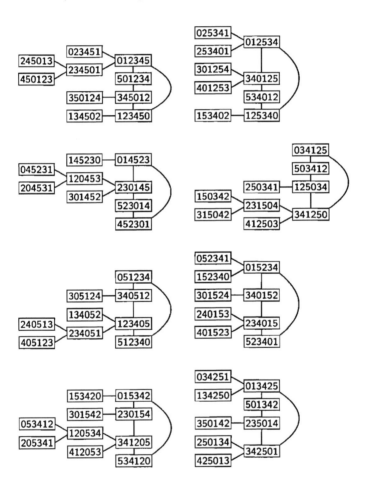

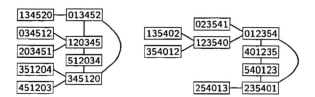

▶ $n = 6,\ e = 12,\ \delta \in \{3, \dots, 5\}$

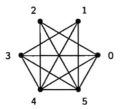

▷ $\gamma = 0.166667\ (m = 1,\ p = 6)$

024135	024153	024315	024513	025134
502413	302415	502431	302451	402513
350241	530241	150243	130245	340251
135024	153024	315024	513024	134025
413502	415302	431502	451302	513402
241350	241530	243150	245130	251340

025143	025314	025413	031425	031452
302514	402531	302541	503142	203145
430251	140253	130254	250314	520314
143025	314025	413025	425031	452031
514302	531402	541302	142503	145203
251430	253140	254130	314250	314520

031524	031542	034152	035142	041352
403152	203154	203415	203514	204135
240315	420315	520341	420351	520413
524031	542031	152034	142035	352041
152403	154203	415203	514203	135204
315240	315420	341520	351420	413520

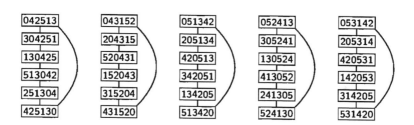

▷ $\gamma = 0.2 \ (m = 1, \ p = 5)$

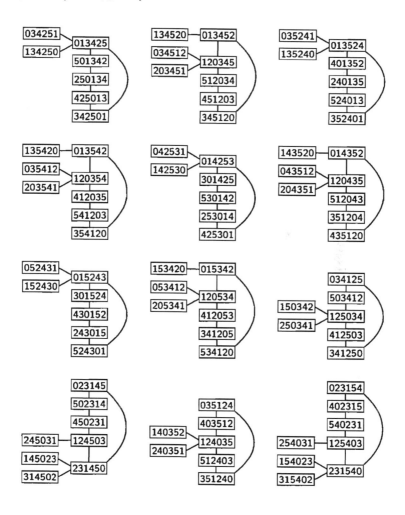

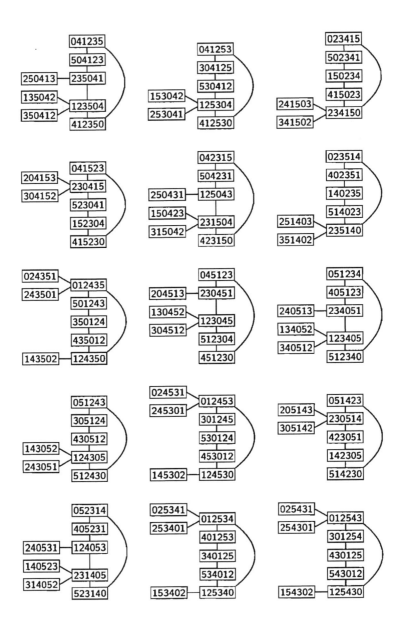

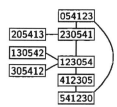

▷ $\gamma = 0.25 \ (m = 1, \ p = 4)$

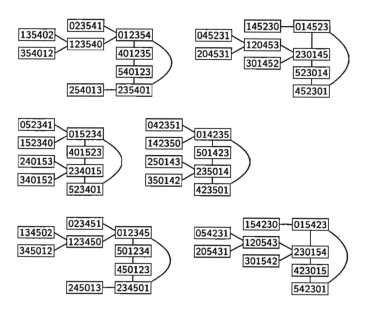

▶ $n = 6,\ e = 10,\ \delta \in \{2, \ldots, 4\}$

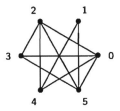

▷ $\gamma = 0.166667\ (m = 1,\ p = 6)$

023514	024153	032415	032514	035142
402351	302415	503241	403251	203514
140235	530241	150324	140325	420351
514023	153024	415032	514032	142035
351402	415302	241503	251403	514203
235140	241530	324150	325140	351420

041523	041532	051423
304152	204153	305142
230415	320415	230514
523041	532041	423051
152304	153204	142305
415230	415320	514230

▷ $\gamma = 0.2\ (m = 1,\ p = 5)$

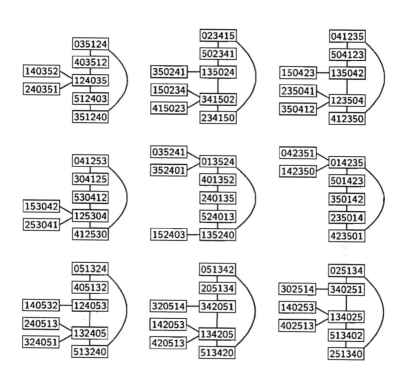

▷ $\gamma = 0.25 \ (m = 1, \ p = 4)$

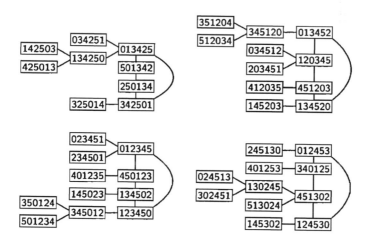

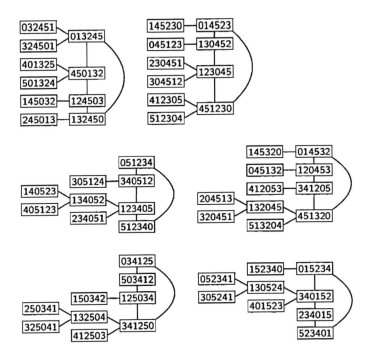

▶ $n = 6,\ e = 10,\ \delta \in \{2, \ldots, 4\}$

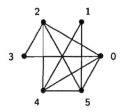

▷ $\gamma = 0.166667\ (m = 1,\ p = 6)$

032415	032514	041523	051423
503241	403251	304152	305142
150324	140325	230415	230514
415032	514032	523041	423051
241503	251403	152304	142305
324150	325140	415230	514230

▷ $\gamma = 0.2\ (m = 1,\ p = 5)$

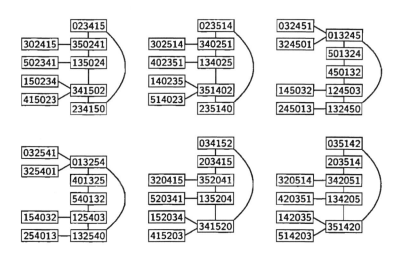

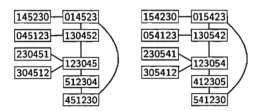

▷ $\gamma = 0.25 \ (m = 1, \ p = 4)$

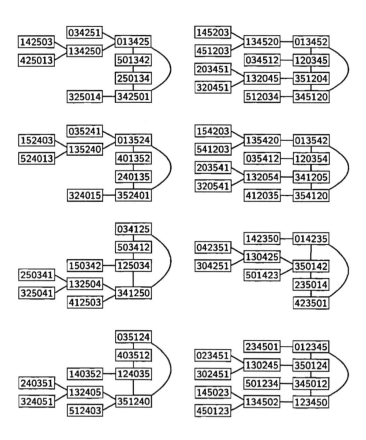

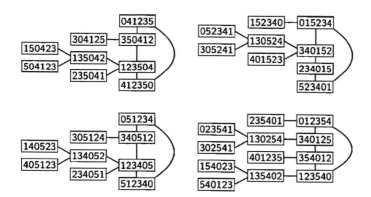

▶ $n = 6$, $e = 11$, $\delta \in \{2, \ldots, 5\}$

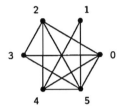

▷ $\gamma = 0.166667$ $(m = 1, p = 6)$

023514	024153	032415	032514	035142
402351	302415	503241	403251	203514
140235	530241	150324	140325	420351
514023	153024	415032	514032	142035
351402	415302	241503	251403	514203
235140	241530	324150	325140	351420

041523	041532	051423
304152	204153	305142
230415	320415	230514
523041	532041	423051
152304	153204	142305
415230	415320	514230

▷ $\gamma = 0.2$ $(m = 1, p = 5)$

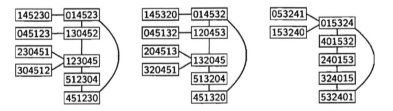

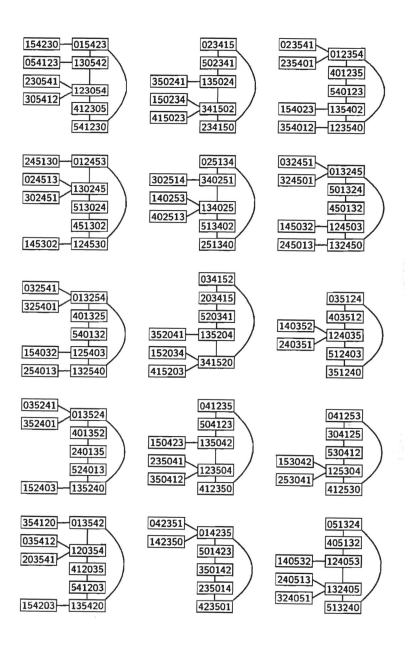

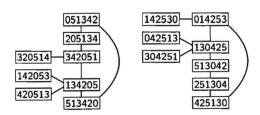

▷ $\gamma = 0.25 \; (m = 1, \; p = 4)$

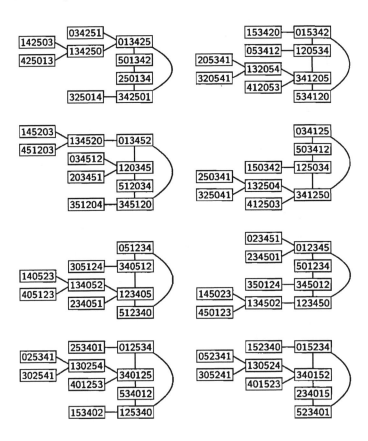

▶ $n = 6$, $e = 10$, $\delta \in \{1, \ldots, 4\}$

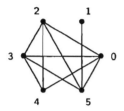

▷ $\gamma = 0.2$ $(m = 1,\ p = 5)$

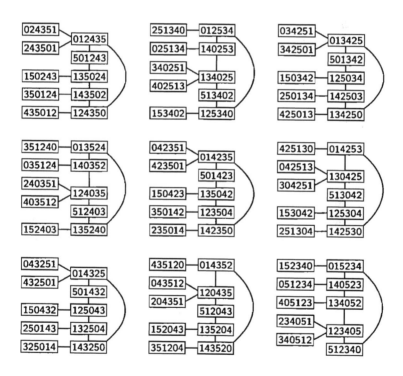

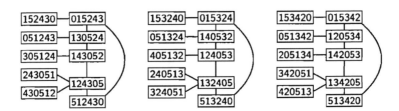

▷ $\gamma = 0.25 \ (m = 1, \ p = 4)$

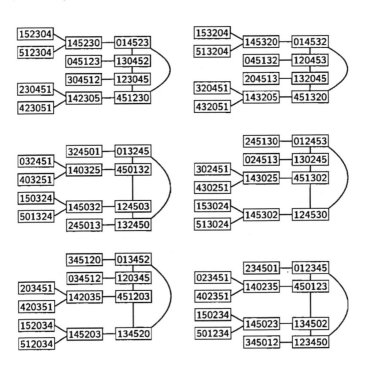

▶ $n = 6$, $e = 11$, $\delta \in \{1, \ldots, 5\}$

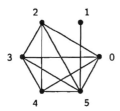

▷ $\gamma = 0.2$ $(m = 1,\, p = 5)$

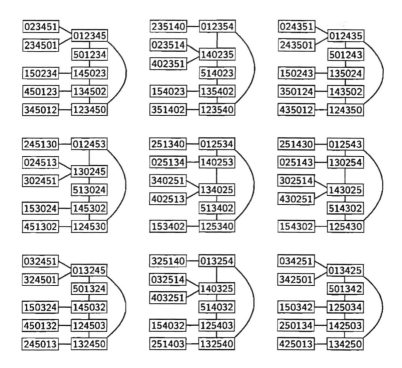

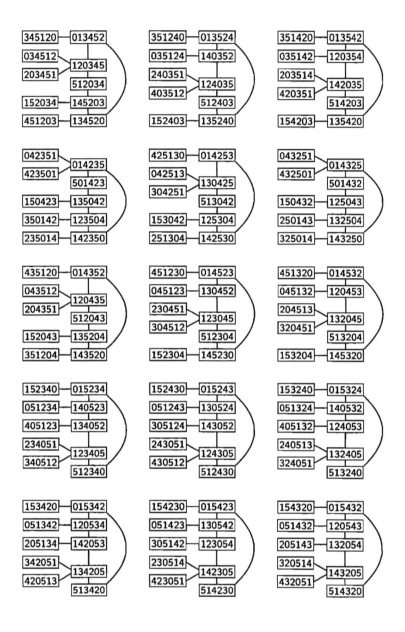

▶ $n = 6$, $e = 12$, $\delta \in \{2, \ldots, 5\}$

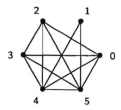

▷ $\gamma = 0.166667$ $(m = 1, p = 6)$

023415	023514	024153	025143	032415
502341	402351	302415	302514	503241
150234	140235	530241	430251	150324
415023	514023	153024	143025	415032
341502	351402	415302	514302	241503
234150	235140	241530	251430	324150

032514	034152	035142	041523	041532
403251	203415	203514	304152	204153
140325	520341	420351	230415	320415
514032	152034	142035	523041	532041
251403	415203	514203	152304	153204
325140	341520	351420	415230	415320

051423	051432
305142	205143
230514	320514
423051	432051
142305	143205
514230	514320

▷ $\gamma = 0.2$ $(m = 1, p = 5)$

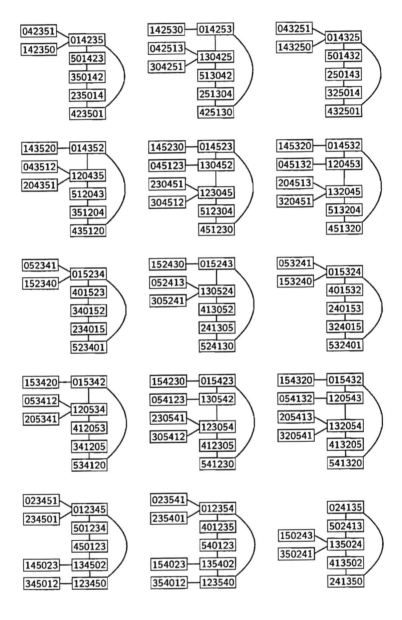

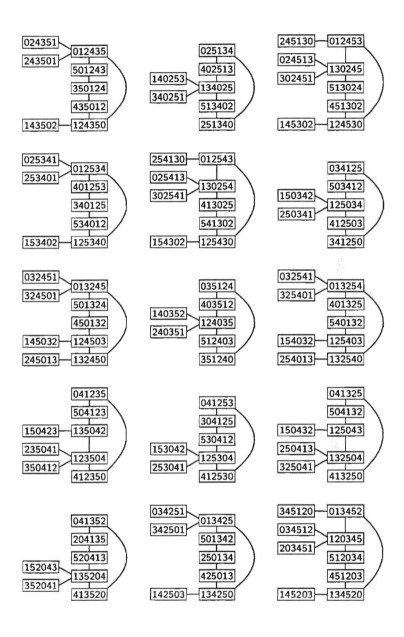

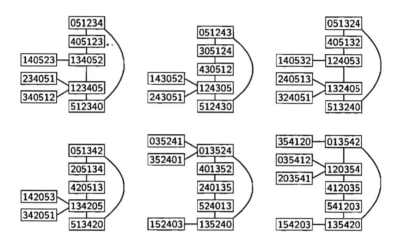

▶ $n = 6$, $e = 13$, $\delta \in \{3, \ldots, 5\}$

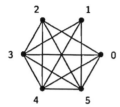

▷ $\gamma = 0.166667$ $(m = 1, p = 6)$

023145	023154	023415	023514	024135
502314	402315	502341	402351	502413
450231	540231	150234	140235	350241
145023	154023	415023	514023	135024
314502	315402	341502	351402	413502
231450	231540	234150	235140	241350

024153	024315	024513	025134	025143
302415	502431	302451	402513	302514
530241	150243	130245	340251	430251
153024	315024	513024	134025	143025
415302	431502	451302	513402	514302
241530	243150	245130	251340	251430

025314	025413	031425	031452	031524
402531	302541	503142	203145	403152
140253	130254	250314	520314	240315
314025	413025	425031	452031	524031
531402	541302	142503	145203	152403
253140	254130	314250	314520	315240

031542	032415	032514	034152	035142
203154	503241	403251	203415	203514
420315	150324	140325	520341	420351
542031	415032	514032	152034	142035
154203	241503	251403	415203	514203
315420	324150	325140	341520	351420

041325	041352	041523	041532	042315
504132	204135	304152	204153	504231
250413	520413	230415	320415	150423
325041	352041	523041	532041	315042
132504	135204	152304	153204	231504
413250	413520	415230	415320	423150

042513	043152	045132	051324	051342
304251	204315	204513	405132	205134
130425	520431	320451	240513	420513
513042	152043	132045	324051	342051
251304	315204	513204	132405	134205
425130	431520	451320	513240	513420

051423	051432	052314	052413	053142
305142	205143	405231	305241	205314
230514	320514	140523	130524	420531
423051	432051	314052	413052	142053
142305	143205	231405	241305	314205
514230	514320	523140	524130	531420

054132
205413
320541
132054
413205
541320

▷ $\gamma = 0.2 \ (m = 1, \ p = 5)$

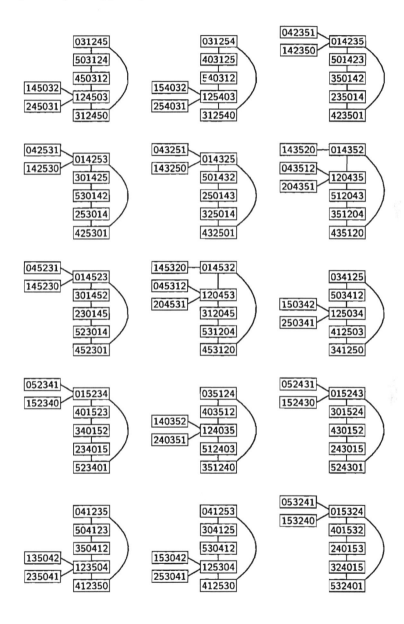

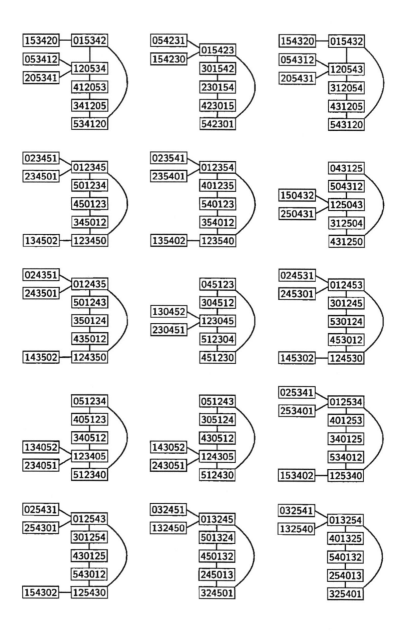

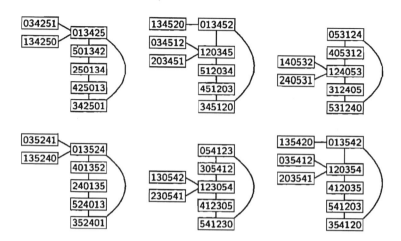

▶ $n = 6, e = 11, \delta \in \{3, 4\}$

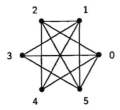

▷ $\gamma = 0.166667\ (m = 1,\ p = 6)$

024135	024153	025314	031425	031524
502413	302415	402531	503142	403152
350241	530241	140253	250314	240315
135024	153024	314025	425031	524031
413502	415302	531402	142503	152403
241350	241530	253140	314250	315240

035124	035142	035214	041253	041352
403512	203514	403521	304125	204135
240351	420351	140352	530412	520413
124035	142035	214035	253041	352041
512403	514203	521403	125304	135204
351240	351420	352140	412530	413520

042135	042153	042513	052413	053124
504213	304215	304251	305241	405312
350421	530421	130425	130524	240531
135042	153042	513042	413052	124053
213504	215304	251304	241305	312405
421350	421530	425130	524130	531240

▷ $\gamma = 0.2$ $(m = 1, p = 5)$

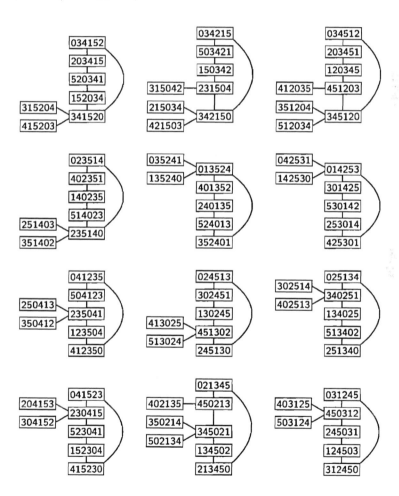

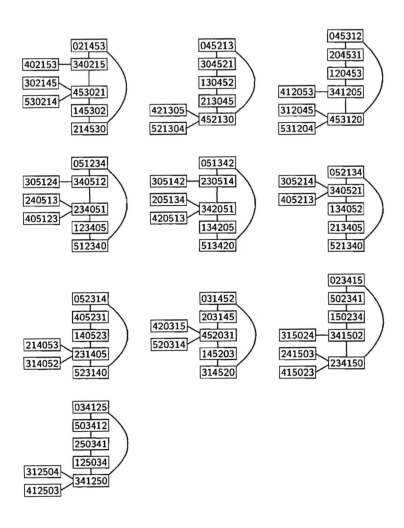

▷ $\gamma = 0.25$ $(m = 1, p = 4)$

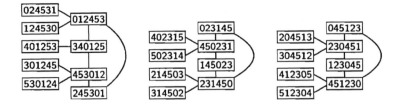

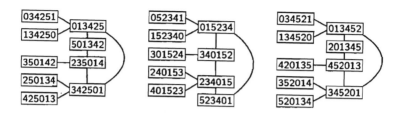

▷ $\gamma = 0.333333$ $(m = 1, p = 3)$

▸ $n = 6,\ e = 11,\ \delta \in \{2, \ldots, 4\}$

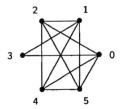

▹ $\gamma = 0.166667\ (m = 1,\ p = 6)$

024513	025413	031245	031254	031425
302451	302541	503124	403125	503142
130245	130254	450312	540312	250314
513024	413025	245031	254031	425031
451302	541302	124503	125403	142503
245130	254130	312450	312540	314250

031452	031524	031542	042513	045213
203145	403152	203154	304251	304521
520314	240315	420315	130425	130452
452031	524031	542031	513042	213045
145203	152403	154203	251304	521304
314520	315240	315420	425130	452130

052413	054213
305241	305421
130524	130542
413052	213054
241305	421305
524130	542130

▷ $\gamma = 0.2 \ (m = 1, \ p = 5)$

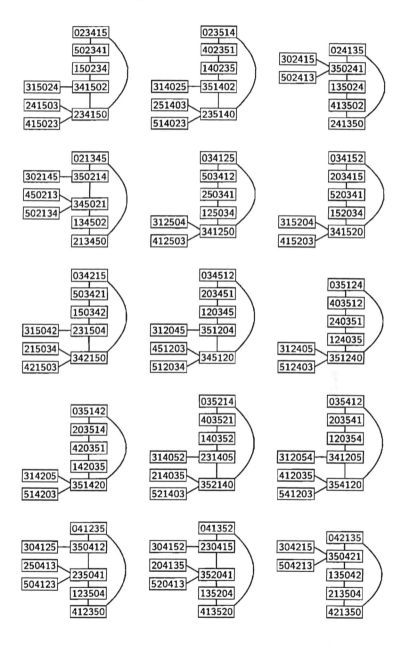

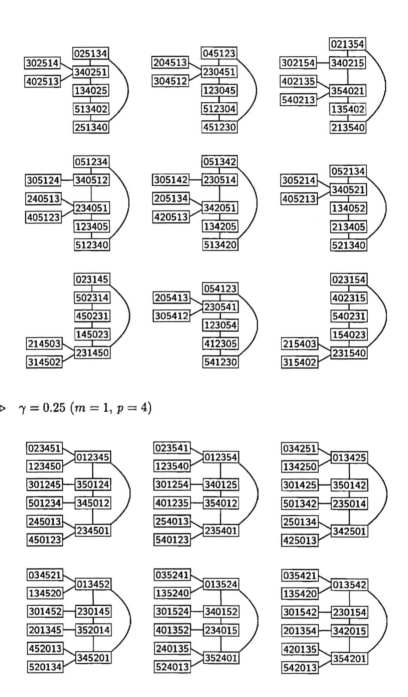

▷ $\gamma = 0.25$ $(m = 1, p = 4)$

▸ $n = 6$, $e = 11$, $\delta \in \{3, 4\}$

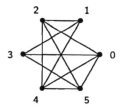

▷ $\gamma = 0.166667$ $(m = 1, p = 6)$

021354	021453	024135	025314	025413
402135	302145	502413	402531	302541
540213	530214	350241	140253	130254
354021	453021	135024	314025	413025
135402	145302	413502	531402	541302
213540	214530	241350	253140	254130

031245	031254	031425	031452	035214
503124	403125	503142	203145	403521
450312	540312	250314	520314	140352
245031	254031	425031	452031	214035
124503	125403	142503	145203	521403
312450	312540	314250	314520	352140

035412	041253	041352	042135	045213
203541	304125	204135	504213	304521
120354	530412	520413	350421	130452
412035	253041	352041	135042	213045
541203	125304	135204	213504	521304
354120	412530	413520	421350	452130

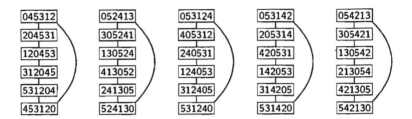

▷ $\gamma = 0.2\ (m = 1,\ p = 5)$

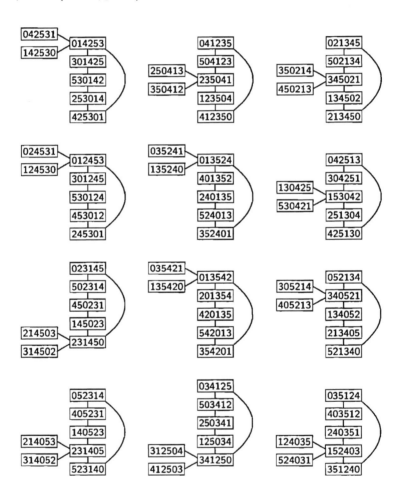

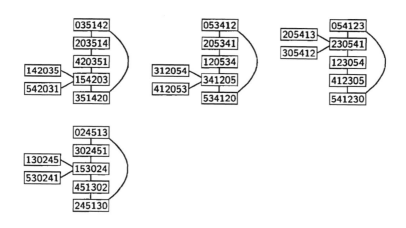

▷ $\gamma = 0.25$ $(m = 1, p = 4)$

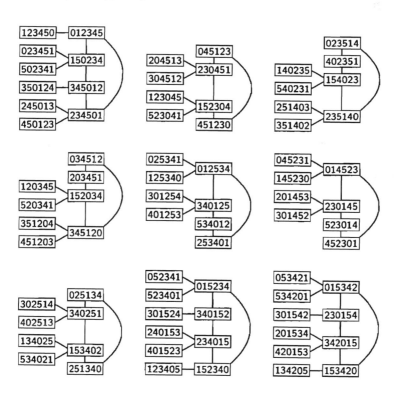

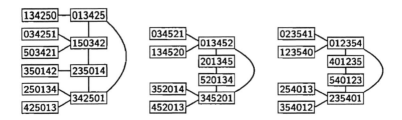

▶ $n = 6,\ e = 12,\ \delta \in \{3, \ldots, 5\}$

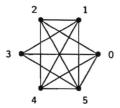

▷ $\gamma = 0.166667\ (m = 1,\ p = 6)$

021354	021453	024135	024153	024513
402135	302145	502413	302415	302451
540213	530214	350241	530241	130245
354021	453021	135024	153024	513024
135402	145302	413502	415302	451302
213540	214530	241350	241530	245130

025314	025413	031245	031254	031425
402531	302541	503124	403125	503142
140253	130254	450312	540312	250314
314025	413025	245031	254031	425031
531402	541302	124503	125403	142503
253140	254130	312450	312540	314250

031452	031524	031542	035124	035142
203145	403152	203154	403512	203514
520314	240315	420315	240351	420351
452031	524031	542031	124035	142035
145203	152403	154203	512403	514203
314520	315240	315420	351240	351420

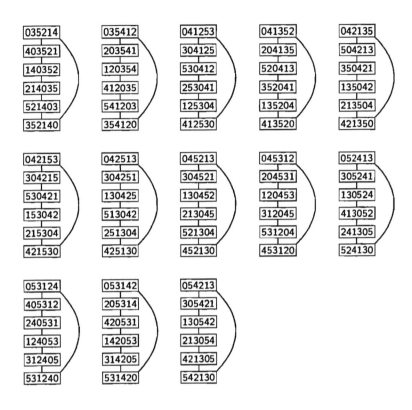

▷ $\gamma = 0.2 \ (m = 1, \ p = 5)$

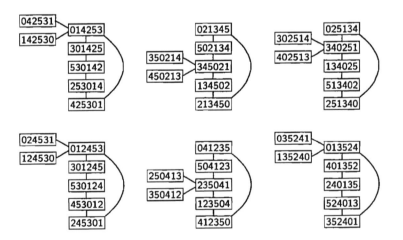

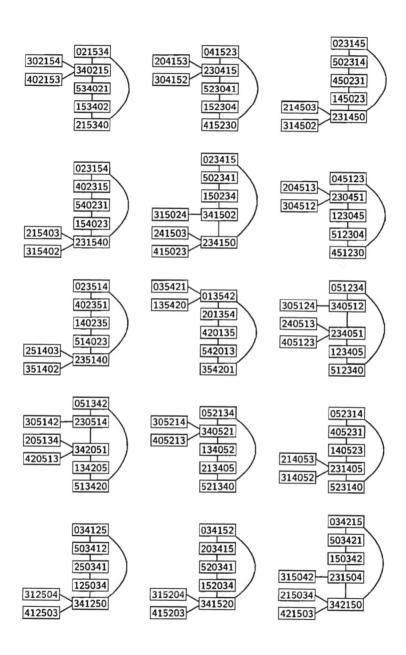

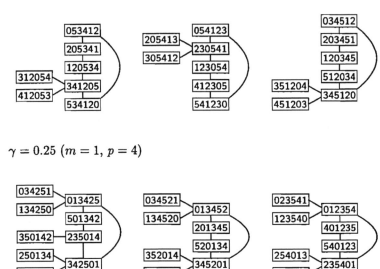

▷ $\gamma = 0.25$ $(m = 1,\ p = 4)$

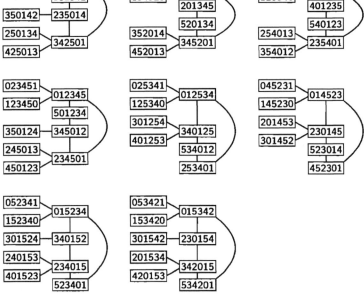

▶ $n = 6$, $e = 12$, $\delta = 4$

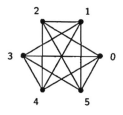

▷ $\gamma = 0.166667$ $(m = 1,\ p = 6)$

021435	021534	024135	024153	024315
502143	402153	502413	302415	502431
350214	340215	350241	530241	150243
435021	534021	135024	153024	315024
143502	153402	413502	415302	431502
214350	215340	241350	241530	243150

025134	025143	025314	031425	031524
402513	302514	402531	503142	403152
340251	430251	140253	250314	240315
134025	143025	314025	425031	524031
513402	514302	531402	142503	152403
251340	251430	253140	314250	315240

034125	034152	034215	035124	035142
503412	203415	503421	403512	203514
250341	520341	150342	240351	420351
125034	152034	215034	124035	142035
412503	415203	421503	512403	514203
341250	341520	342150	351240	351420

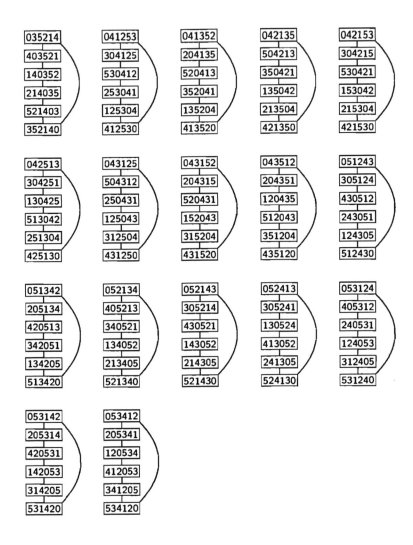

▷ $\gamma = 0.2$ $(m = 1, p = 5)$

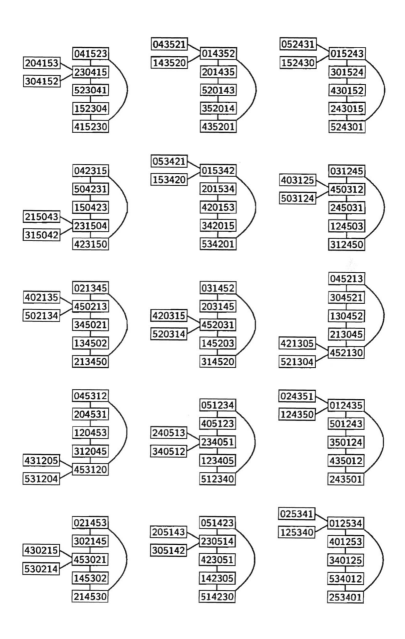

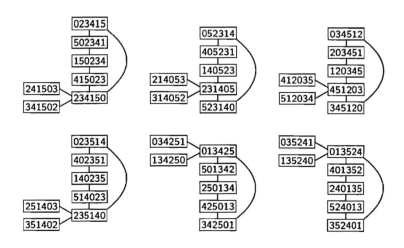

▷ $\gamma = 0.25$ $(m = 1, p = 4)$

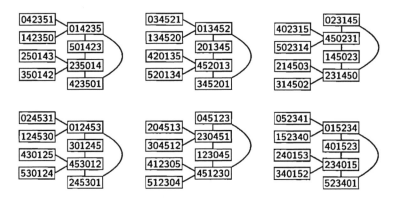

▷ $\gamma = 0.333333$ $(m = 1, p = 3)$

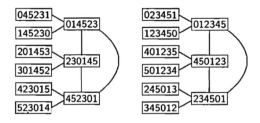

▶ $n = 6$, $e = 13$, $\delta \in \{4, 5\}$

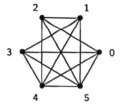

▷ $\gamma = 0.166667$ $(m = 1, p = 6)$

021345	021354	021435	021453	021534
502134	402135	502143	302145	402153
450213	540213	350214	530214	340215
345021	354021	435021	453021	534021
134502	135402	143502	145302	153402
213450	213540	214350	214530	215340

021543	024135	024153	024315	024513
302154	502413	302415	502431	302451
430215	350241	530241	150243	130245
543021	135024	153024	315024	513024
154302	413502	415302	431502	451302
215430	241350	241530	243150	245130

025134	025143	025314	025413	031245
402513	302514	402531	302541	503124
340251	430251	140253	130254	450312
134025	143025	314025	413025	245031
513402	514302	531402	541302	124503
251340	251430	253140	254130	312450

031254	031425	031452	031524	031542
403125	503142	203145	403152	203154
540312	250314	520314	240315	420315
254031	425031	452031	524031	542031
125403	142503	145203	152403	154203
312540	314250	314520	315240	315420

034125	034152	034215	034512	035124
503412	203415	503421	203451	403512
250341	520341	150342	120345	240351
125034	152034	215034	512034	124035
412503	415203	421503	451203	512403
341250	341520	342150	345120	351240

035142	035214	035412	041253	041352
203514	403521	203541	304125	204135
420351	140352	120354	530412	520413
142035	214035	412035	253041	352041
514203	521403	541203	125304	135204
351420	352140	354120	412530	413520

042135	042153	042513	043125	043152
504213	304215	304251	504312	204315
350421	530421	130425	250431	520431
135042	153042	513042	125043	152043
213504	215304	251304	312504	315204
421350	421530	425130	431250	431520

043512	045213	045312	051243	051342
204351	304521	204531	305124	205134
120435	130452	120453	430512	420513
512043	213045	312045	243051	342051
351204	521304	531204	124305	134205
435120	452130	453120	512430	513420

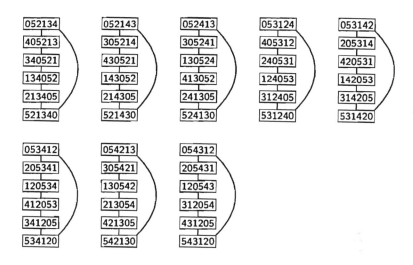

▷ $\gamma = 0.2 \ (m = 1, \ p = 5)$

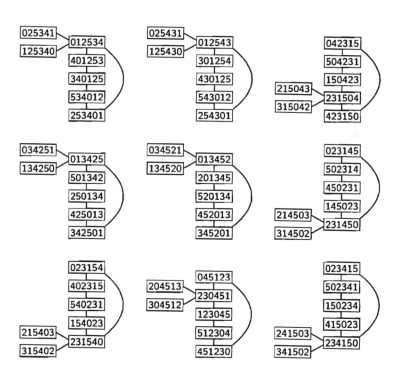

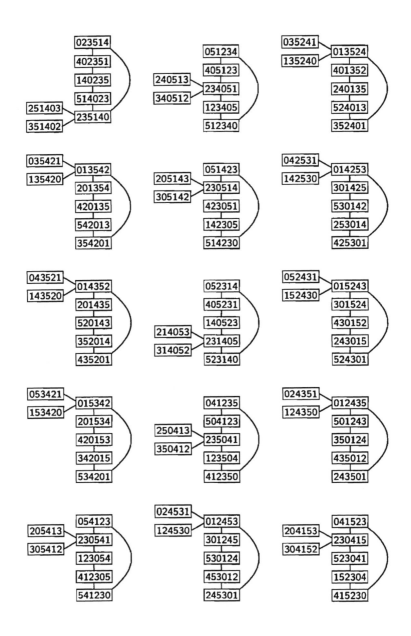

▷ $\gamma = 0.25 \ (m = 1, \ p = 4)$

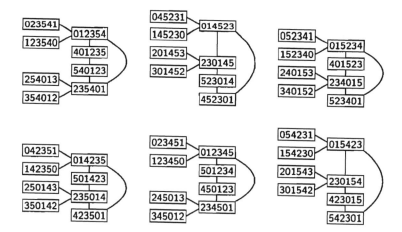

▶ $n = 6$, $e = 14$, $\delta \in \{4, 5\}$

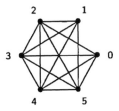

▷ $\gamma = 0.166667$ $(m = 1, p = 6)$

021345	021354	021435	021453	021534
502134	402135	502143	302145	402153
450213	540213	350214	530214	340215
345021	354021	435021	453021	534021
134502	135402	143502	145302	153402
213450	213540	214350	214530	215340

021543	023145	023154	023415	023514
302154	502314	402315	502341	402351
430215	450231	540231	150234	140235
543021	145023	154023	415023	514023
154302	314502	315402	341502	351402
215430	231450	231540	234150	235140

024135	024153	024315	024513	025134
502413	302415	502431	302451	402513
350241	530241	150243	130245	340251
135024	153024	315024	513024	134025
413502	415302	431502	451302	513402
241350	241530	243150	245130	251340

025143	025314	025413	031245	031254
302514	402531	302541	503124	403125
430251	140253	130254	450312	540312
143025	314025	413025	245031	254031
514302	531402	541302	124503	125403
251430	253140	254130	312450	312540

031425	031452	031524	031542	032145
503142	203145	403152	203154	503214
250314	520314	240315	420315	450321
425031	452031	524031	542031	145032
142503	145203	152403	154203	214503
314250	314520	315240	315420	321450

032154	032415	032514	034125	034152
403215	503241	403251	503412	203415
540321	150324	140325	250341	520341
154032	415032	514032	125034	152034
215403	241503	251403	412503	415203
321540	324150	325140	341250	341520

034215	034512	035124	035142	035214
503421	203451	403512	203514	403521
150342	120345	240351	420351	140352
215034	512034	124035	142035	214035
421503	451203	512403	514203	521403
342150	345120	351240	351420	352140

035412	041235	041253	041325	041352
203541	504123	304125	504132	204135
120354	350412	530412	250413	520413
412035	235041	253041	325041	352041
541203	123504	125304	132504	135204
354120	412350	412530	413250	413520

041523	041532	042135	042153	042315
304152	204153	504213	304215	504231
230415	320415	350421	530421	150423
523041	532041	135042	153042	315042
152304	153204	213504	215304	231504
415230	415320	421350	421530	423150

042513	043125	043152	043215	043512
304251	504312	204315	504321	204351
130425	250431	520431	150432	120435
513042	125043	152043	215043	512043
251304	312504	315204	321504	351204
425130	431250	431520	432150	435120

045123	045132	045213	045312	051234
304512	204513	304521	204531	405123
230451	320451	130452	120453	340512
123045	132045	213045	312045	234051
512304	513204	521304	531204	123405
451230	451320	452130	453120	512340

051243	051324	051342	051423	051432
305124	405132	205134	305142	205143
430512	240513	420513	230514	320514
243051	324051	342051	423051	432051
124305	132405	134205	142305	143205
512430	513240	513420	514230	514320

052134	052143	052314	052413	053124
405213	305214	405231	305241	405312
340521	430521	140523	130524	240531
134052	143052	314052	413052	124053
213405	214305	231405	241305	312405
521340	521430	523140	524130	531240

053142
205314
420531
142053
314205
531420

053214
405321
140532
214053
321405
532140

053412
205341
120534
412053
341205
534120

054123
305412
230541
123054
412305
541230

054132
205413
320541
132054
413205
541320

054213
305421
130542
213054
421305
542130

054312
205431
120543
312054
431205
543120

▷ $\gamma = 0.2\ (m = 1,\ p = 5)$

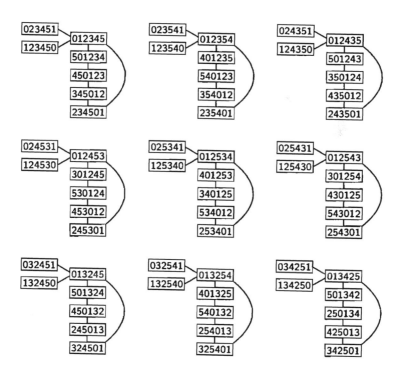

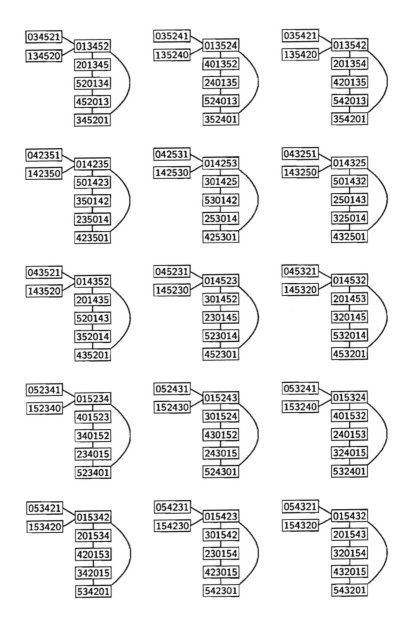

▶ $n = 6$, $e = 15$, $\delta = 5$

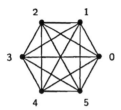

▷ $\gamma = 0.166667$ $(m = 1, p = 6)$

012345	012354	012435	012453	012534
501234	401235	501243	301245	401253
450123	540123	350124	530124	340125
345012	354012	435012	453012	534012
234501	235401	243501	245301	253401
123450	123540	124350	124530	125340

012543	013245	013254	013425	013452
301254	501324	401325	501342	201345
430125	450132	540132	250134	520134
543012	245013	254013	425013	452013
254301	324501	325401	342501	345201
125430	132450	132540	134250	134520

013524	013542	014235	014253	014325
401352	201354	501423	301425	501432
240135	420135	350142	530142	250143
524013	542013	235014	253014	325014
352401	354201	423501	425301	432501
135240	135420	142350	142530	143250

014352	014523	014532	015234	015243
201435	301452	201453	401523	301524
520143	230145	320145	340152	430152
352014	523014	532014	234015	243015
435201	452301	453201	523401	524301
143520	145230	145320	152340	152430

015324	015342	015423	015432	021345
401532	201534	301542	201543	502134
240153	420153	230154	320154	450213
324015	342015	423015	432015	345021
532401	534201	542301	543201	134502
153240	153420	154230	154320	213450

021354	021435	021453	021534	021543
402135	502143	302145	402153	302154
540213	350214	530214	340215	430215
354021	435021	453021	534021	543021
135402	143502	145302	153402	154302
213540	214350	214530	215340	215430

023145	023154	023415	023451	023514
502314	402315	502341	102345	402351
450231	540231	150234	510234	140235
145023	154023	415023	451023	514023
314502	315402	341502	345102	351402
231450	231540	234150	234510	235140

023541	024135	024153	024315	024351
102354	502413	302415	502431	102435
410235	350241	530241	150243	510243
541023	135024	153024	315024	351024
354102	413502	415302	431502	435102
235410	241350	241530	243150	243510

024513	024531	025134	025143	025314
302451	102453	402513	302514	402531
130245	310245	340251	430251	140253
513024	531024	134025	143025	314025
451302	453102	513402	514302	531402
245130	245310	251340	251430	253140

025341	025413	025431	031245	031254
102534	302541	102543	503124	403125
410253	130254	310254	450312	540312
341025	413025	431025	245031	254031
534102	541302	543102	124503	125403
253410	254130	254310	312450	312540

031425	031452	031524	031542	032145
503142	203145	403152	203154	503214
250314	520314	240315	420315	450321
425031	452031	524031	542031	145032
142503	145203	152403	154203	214503
314250	314520	315240	315420	321450

032154	032415	032451	032514	032541
403215	503241	103245	403251	103254
540321	150324	510324	140325	410325
154032	415032	451032	514032	541032
215403	241503	245103	251403	254103
321540	324150	324510	325140	325410

034125	034152	034215	034251	034512
503412	203415	503421	103425	203451
250341	520341	150342	510342	120345
125034	152034	215034	251034	512034
412503	415203	421503	425103	451203
341250	341520	342150	342510	345120

034521	035124	035142	035214	035241
103452	403512	203514	403521	103524
210345	240351	420351	140352	410352
521034	124035	142035	214035	241035
452103	512403	514203	521403	524103
345210	351240	351420	352140	352410

035412	035421	041235	041253	041325
203541	103542	504123	304125	504132
120354	210354	350412	530412	250413
412035	421035	235041	253041	325041
541203	542103	123504	125304	132504
354120	354210	412350	412530	413250

041352	041523	041532	042135	042153
204135	304152	204153	504213	304215
520413	230415	320415	350421	530421
352041	523041	532041	135042	153042
135204	152304	153204	213504	215304
413520	415230	415320	421350	421530

042315	042351	042513	042531	043125
504231	104235	304251	104253	504312
150423	510423	130425	310425	250431
315042	351042	513042	531042	125043
231504	235104	251304	253104	312504
423150	423510	425130	425310	431250

043152	043215	043251	043512	043521
204315	504321	104325	204351	104352
520431	150432	510432	120435	210435
152043	215043	251043	512043	521043
315204	321504	325104	351204	352104
431520	432150	432510	435120	435210

045123	045132	045213	045231	045312
304512	204513	304521	104523	204531
230451	320451	130452	310452	120453
123045	132045	213045	231045	312045
512304	513204	521304	523104	531204
451230	451320	452130	452310	453120

045321	051234	051243	051324	051342
104532	405123	305124	405132	205134
210453	340512	430512	240513	420513
321045	234051	243051	324051	342051
532104	123405	124305	132405	134205
453210	512340	512430	513240	513420

051423	051432	052134	052143	052314
305142	205143	405213	305214	405231
230514	320514	340521	430521	140523
423051	432051	134052	143052	314052
142305	143205	213405	214305	231405
514230	514320	521340	521430	523140

052341	052413	052431	053124	053142
105234	305241	105243	405312	205314
410523	130524	310524	240531	420531
341052	413052	431052	124053	142053
234105	241305	243105	312405	314205
523410	524130	524310	531240	531420

053214	053241	053412	053421	054123
405321	105324	205341	105342	305412
140532	410532	120534	210534	230541
214053	241053	412053	421053	123054
321405	324105	341205	342105	412305
532140	532410	534120	534210	541230

054132	054213	054231	054312	054321
205413	305421	105423	205431	105432
320541	130542	310542	120543	210543
132054	213054	231054	312054	321054
413205	421305	423105	431205	432105
541320	542130	542310	543120	543210

6

All Trees on Seven Nodes

This chapter is the atlas of SER basins of attraction for graphs in $\mathcal{T}_7$. It contains entries for all the 11 trees with $n = 7$.

▶ $n = 7$, $e = 6$, $\delta \in \{1, \ldots, 6\}$

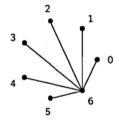

▷ $\gamma = 0.5\ (m = 1,\ p = 2)$

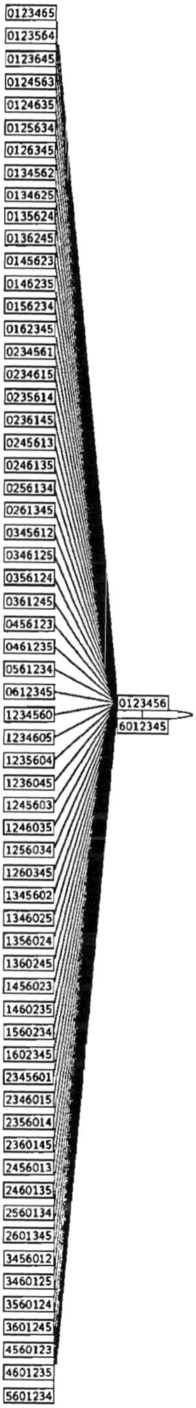

0123465
0123564
0123645
0124563
0124635
0125634
0126345
0134562
0134625
0135624
0136245
0145623
0146235
0156234
0162345
0234561
0234615
0235614
0236145
0245613
0246135
0256134
0261345
0345612
0346125
0356124
0361245
0456123
0461235
0561234
0612345
1234560
1234605
1235604
1236045
1245603
1246035
1256034
1260345
1345602
1346025
1356024
1360245
1456023
1460235
1560234
1602345
2345601
2346015
2356014
2360145
2456013
2460135
2560134
2601345
3456012
3460125
3560124
3601245
4560123
4601235
5601234

0123456
6012345

▶ $n = 7,\ e = 6,\ \delta \in \{1,\dots,5\}$

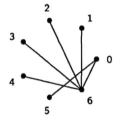

$\triangleright$ $\gamma = 0.5$ $(m = 1, \, p = 2)$

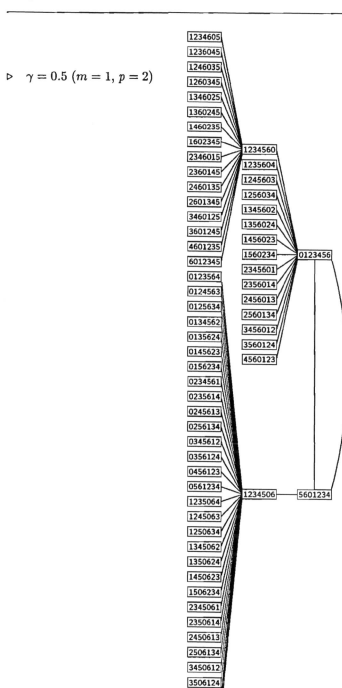

▶ $n = 7, e = 6, \delta \in \{1, \ldots, 4\}$

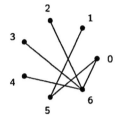

▷ $\gamma = 0.5$ $(m = 1,\, p = 2)$

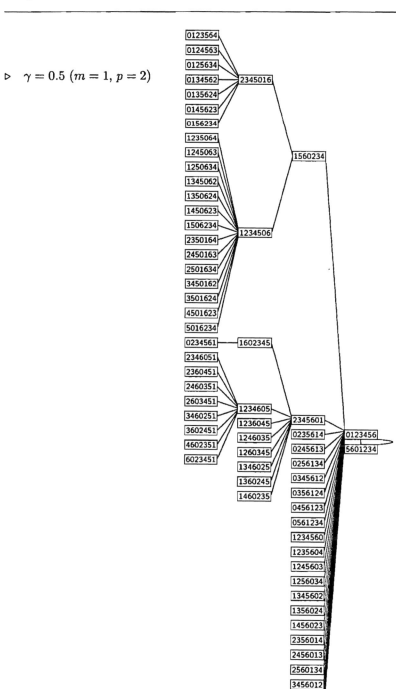

▶ $n = 7,\ e = 6,\ \delta \in \{1, \ldots, 4\}$

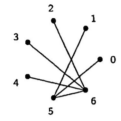

▷ $\gamma = 0.5\ (m = 1,\ p = 2)$

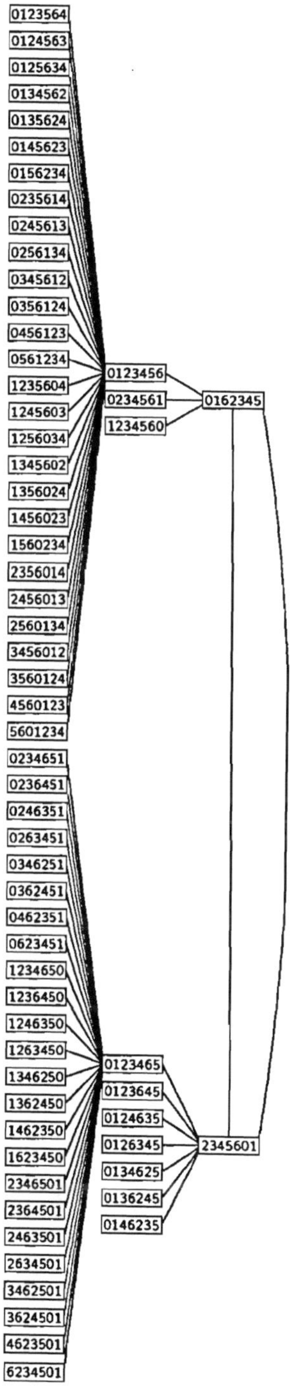

▶ $n = 7$, $e = 6$, $\delta \in \{1, \ldots, 3\}$

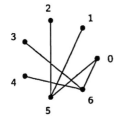

▷ $\gamma = 0.5\ (m = 1,\ p = 2)$

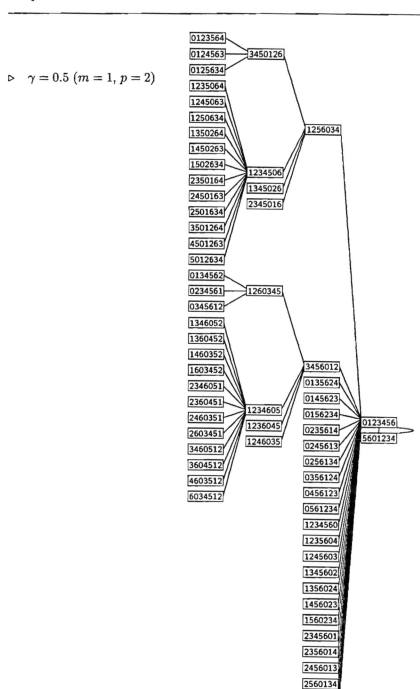

▸ $n = 7, e = 6, \delta \in \{1, \ldots, 4\}$

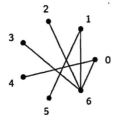

▷ $\gamma = 0.5$ $(m = 1,\ p = 2)$

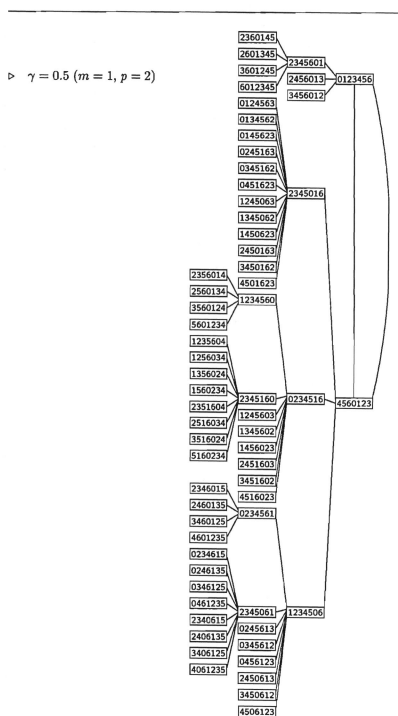

▶ $n = 7, e = 6, \delta \in \{1, \ldots, 3\}$

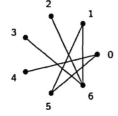

▷ $\gamma = 0.5$ ($m = 1$, $p \approx 2$)

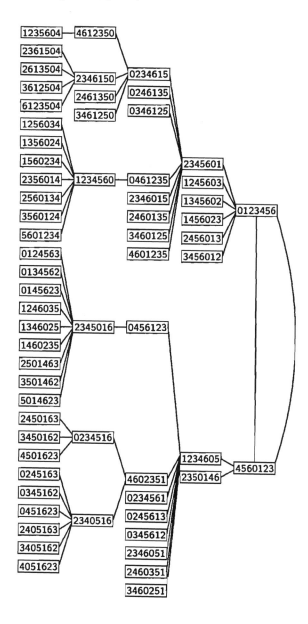

▶ $n = 7,\, e = 6,\, \delta \in \{1, \ldots, 3\}$

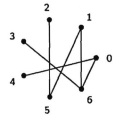

▷ $\gamma = 0.5$ $(m = 1, p = 2)$

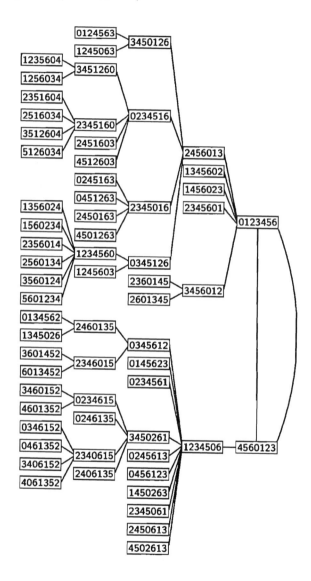

▶ $n = 7,\ e = 6,\ \delta \in \{1,\ldots,3\}$

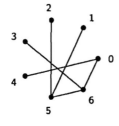

▷ $\gamma = 0.5$ $(m = 1, p = 2)$

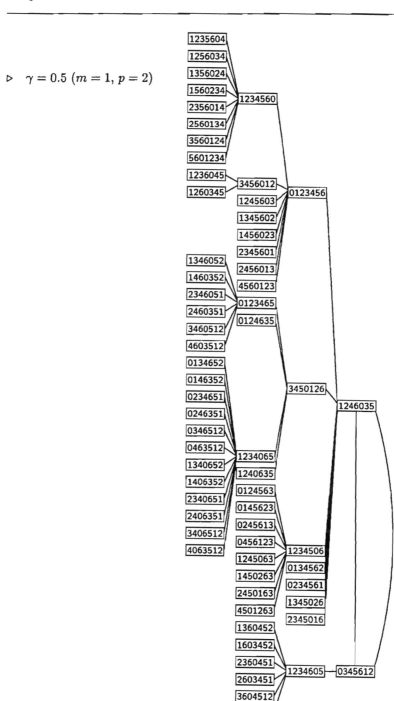

▶ $n = 7$, $e = 6$, $\delta \in \{1, \ldots, 3\}$

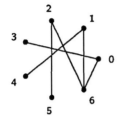

▷ $\gamma = 0.5 \ (m = 1, \ p = 2)$

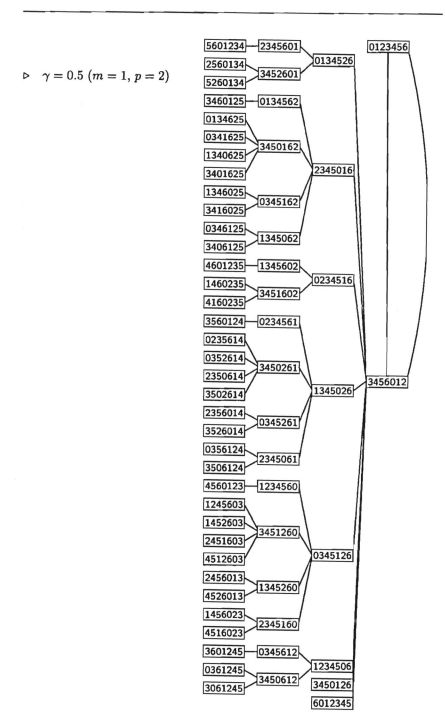

▶ $n = 7$, $e = 6$, $\delta \in \{1, 2\}$

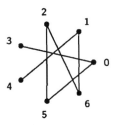

▷ $\gamma = 0.5$ $(m = 1,\ p = 2)$

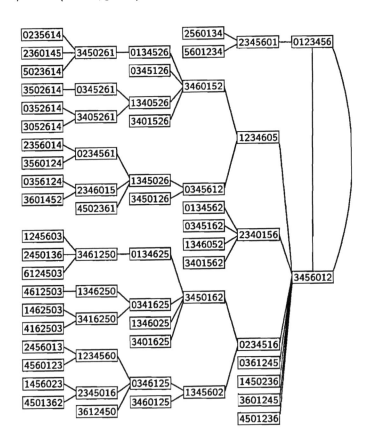

7

All Rings on up to Eight Nodes

This chapter is the atlas of SER basins of attraction for graphs in $\mathcal{R}_3$ through $\mathcal{R}_8$. It contains entries for all the 6 rings for which $3 \leq n \leq 8$.

▶ $n = 3$, $e = 3$, $\delta = 2$

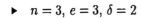

▷ $\gamma = 0.333333$ $(m = 1, p = 3)$

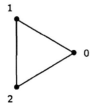

▶ $n = 4$, $e = 4$, $\delta = 2$

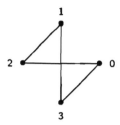

▷ $\gamma = 0.25$ $(m = 1, p = 4)$

▷ $\gamma = 0.5$ $(m = 1, p = 2)$

0231	
1230	0123
2013	
3012	2301

▶ $n = 5$, $e = 5$, $\delta = 2$

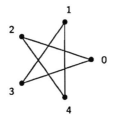

▷ $\gamma = 0.2$ $(m = 1, p = 5)$

02413		03142
30241		20314
13024		42031
41302		14203
24130		31420

▷ $\gamma = 0.4$ $(m = 2, p = 5)$

02341—01234		13420—01342
30124—34012		30142—23014
13402—12340		02314—03412
41230—04123		24031—12403
24013—23401		41203—34120

▶ $n = 6$, $e = 6$, $\delta = 2$

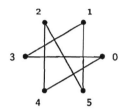

▷ $\gamma = 0.166667$ $(m = 1, p = 6)$

031524	042513
403152	304251
240315	130425
524031	513042
152403	251304
315240	425130

▷ $\gamma = 0.333333$ $(m = 2, p = 6)$

034251—013425	135240—013524
301425—350142	301524—340152
135042—123504	034152—023415
512304—451230	402315—450231
245130—024513	245031—124503
402513—340251	512403—351240

▷ $\gamma = 0.333333$ $(m = 1, p = 3)$

035124	045123
145023	235014
234015	134025

▷ $\gamma = 0.5 \ (m = 1, \ p = 2)$

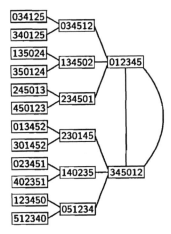

▸ $n = 7,\, e = 7,\, \delta = 2$

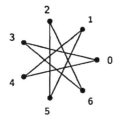

▷ $\gamma = 0.142857\ (m = 1,\, p = 7)$

0362514	0415263
4036251	3041526
1403625	6304152
5140362	2630415
2514036	5263041
6251403	1526304
3625140	4152630

▷ $\gamma = 0.285714\ (m = 2,\, p = 7)$

0235614	0246135	0346251 — 0134625
4602351	3502461	4013625 — 4501362
1346025	1562304	1450362 — 1245036
0356124	2461350	5124036 — 5612403
2450136	0451263	2561403 — 2356140
1460235	1350246	6235140 — 0623514
3561240	4613502	3602514 — 3460251

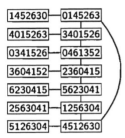

▷ $\gamma = 0.428571$ $(m = 3,\ p = 7)$

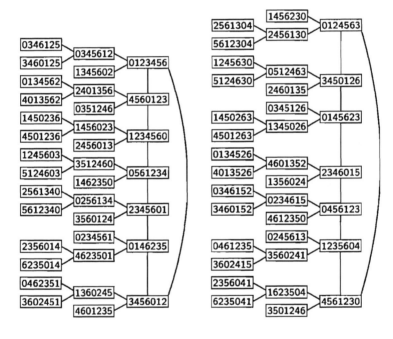

▶ $n = 8, e = 8, \delta = 2$

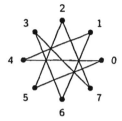

▷ $\gamma = 0.125 \ (m = 1, p = 8)$

04163725	05273614
50416372	40527361
25041637	14052736
72504163	61405273
37250416	36140527
63725041	73614052
16372504	27361405
41637250	52736140

▷ $\gamma = 0.25 \ (m = 2, p = 8)$

04527361—01452736	04613725	14637250—01463725
40152736—46015273	15603472	40163725—45016372
14605273—13460527	24501637	04516372—02451637
61340527—67134052	04712563	50241637—57024163
36714052—23671405	35702416	25704163—23570416
72361405—57236140	25603471	72350416—67235041
25736140—02573614	16723504	36725041—13672504
50273614—45027361	34712560	61372504—46137250

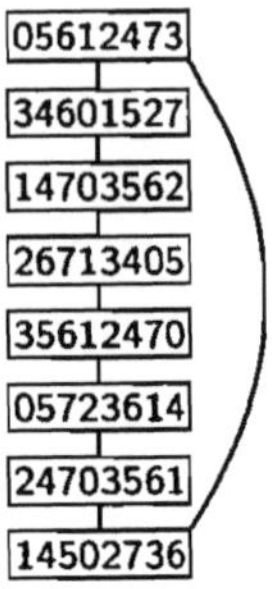

05612473
34601527
14703562
26713405
35612470
05723614
24703561
14502736

▷ $\gamma = 0.25 \ (m = 1, \ p = 4)$

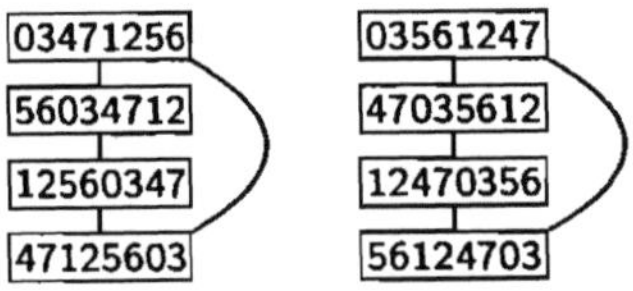

03471256   03561247
56034712   47035612
12560347   12470356
47125603   56124703

▷ $\gamma = 0.375 \ (m = 3, \ p = 8)$

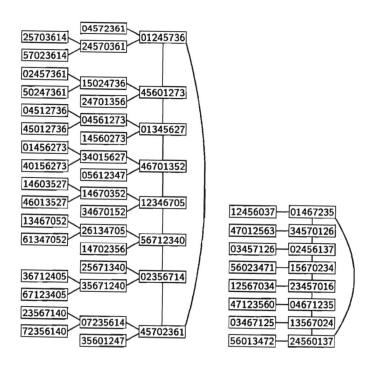

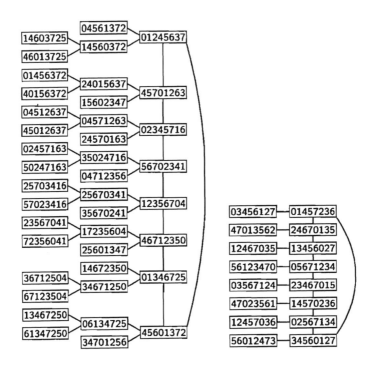

▷ $\gamma = 0.5 \ (m = 1, \ p = 2)$

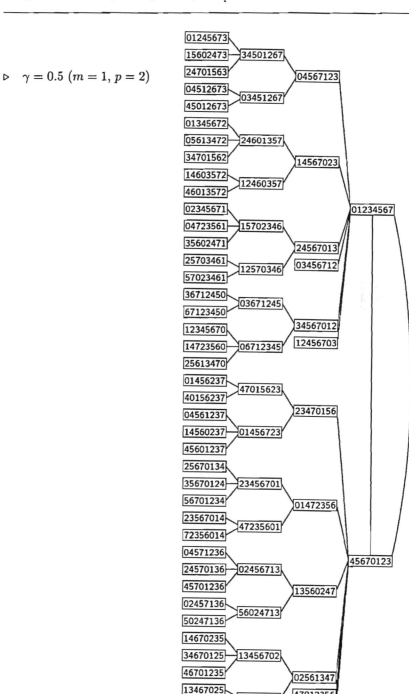

References

1. Attiya, H., H. Shachnai, and T. Tamir (1999). Local labeling and resource allocation using preprocessing. *SIAM J. on Computing* **28**, 1397–1413.
2. Awerbuch, B. (1985). Complexity of network synchronization. *J. of the ACM* **32**, 804–823.
3. Barbosa, V. C. (1986). Concurrency in systems with neighborhood constraints. Ph.D. dissertation, Computer Science Department, University of California, Los Angeles, CA.
4. Barbosa, V. C. (1993). *Massively Parallel Models of Computation: Distributed Parallel Processing in Artificial Intelligence and Optimization.* Ellis Horwood, Chichester, UK.
5. Barbosa, V. C. (1996). *An Introduction to Distributed Algorithms.* The MIT Press, Cambridge, MA.
6. Barbosa, V. C., M. R. F. Benevides, and F. M. G. França (2000). Sharing resources at nonuniform access rates. *Theory of Computing Systems*, to appear.
7. Barbosa, V. C., and E. Gafni (1989). A distributed implementation of simulated annealing. *J. of Parallel and Distributed Computing* **6**, 411–434.
8. Barbosa, V. C., and E. Gafni (1989). Concurrency in heavily loaded neighborhood-constrained systems. *ACM Trans. on Programming Languages and Systems* **11**, 562–584.

9. Barbosa, V. C., and P. M. V. Lima (1990). On the distributed parallel simulation of Hopfield's neural networks. *Software—Practice and Experience* **20**, 967–983.

10. Barbosa, V. C., and J. L. Szwarcfiter (1999). Generating all the acyclic orientations of an undirected graph. *Information Processing Letters* **72**, 71–74.

11. Berge, C. (1976). *Graphs and Hypergraphs.* North-Holland, Amsterdam, The Netherlands.

12. Bertsekas, D. P., and J. N. Tsitsiklis (1989). *Parallel and Distributed Computation: Numerical Methods.* Prentice-Hall, Englewood Cliffs, NJ.

13. Bertsekas, D. P., and J. N. Tsitsiklis (1991). Some aspects of parallel and distributed iterative algorithms—a survey. *Automatica* **27**, 3–21.

14. Besag, J. (1974). Spatial interaction and the statistical analysis of lattice systems. *J. of the Royal Statistical Society, Series B* **36**, 192–236.

15. Bondy, J. A., and U. S. R. Murty (1976). *Graph Theory with Applications.* North-Holland, New York, NY.

16. Brilman, M., and J.-M. Vincent (1998). On the estimation of the throughput for a class of stochastic resources sharing systems. *Mathematics of Operations Research* **23**, 305–321.

17. Calabrese, A., and F. M. G. França (1994). A randomised distributed primer for the updating control of ANNs. In *Proc. of the International Conference on Artificial Neural Networks*, 585–588. Springer-Verlag, New York, NY.

18. Cameron, R. D., C. J. Colbourn, R. C. Read, and N. C. Wormald (1985). Cataloguing the graphs on 10 vertices. *J. of Graph Theory* **9**, 551–562.

19. Chandy, K. M., and J. Misra (1984). The drinking philosophers problem. *ACM Trans. on Programming Languages and Systems* **6**, 632–646.

20. Chandy, K. M., and J. Misra (1988). *Parallel Program Design: a Foundation.* Addison-Wesley, Reading, MA.

21. Clarke, F. H., and R. E. Jamison (1976). Multicolorings, measures and games on graphs. *Discrete Mathematics* **14**, 241–245.

22. Cooper, G. F. (1990). The computational complexity of probabilistic inference using Bayesian belief networks. *Artificial Intelligence* **42**, 393–405.

23. d'Acierno, A., and R. Vaccaro (1994). On parallelizing recursive neural networks on coarse-grained parallel computers: a general algorithm. *Parallel Computing* **20**, 245–256.

24. Dijkstra, E. W. (1971). Hierarchical ordering of sequential processes. *Acta Informatica* **1**, 115–138.

25. Eizirik, L. M. R., V. C. Barbosa, and S. B. T. Mendes (1993). A Bayesian-network approach to lexical disambiguation. *Cognitive Science* **17**, 257–283.

26. Even, S., and S. Rajsbaum (1997). The use of a synchronizer yields the maximum computation rate in distributed networks. *Theory of Computing Systems* **30**, 447–474.

27. Felten, E., S. Karlin, and S. W. Otto (1985). The traveling salesman problem on a hypercubic, MIMD computer. In *Proc. of the International Conference on Parallel Processing*, 6–10. IEEE Computer Society Press, Washington, DC.

28. França, F. M. G. (1991). A self-organizing updating network. In *Proc. of the International Conference on Artificial Neural Networks*, 1349–1352. North-Holland, New York, NY.

29. França, F. M. G. (1992). Distributed generation of synchronous, asynchronous and partially synchronous/asynchronous updating dynamics by a self-organizing oscillatory network. In *Proc. of the International Conference on Artificial Neural Networks*, 1057–1061. North-Holland, New York, NY.

30. França, F. M. G. (1993). Scheduling weightless systems with self-timed Boolean networks. In *Proc. of the Workshop on Weightless Neural Networks*, 87–92.

31. França, F. M. G. (1994). Neural networks as neighbourhood-constrained systems. Ph.D. thesis, Department of Electrical and Electronic Engineering, Imperial College of Science, Technology, and Medicine, London, UK.

32. Gafni, E. M., and D. P. Bertsekas (1981). Distributed algorithms for generating loop-free routes in networks with frequently changing topology. *IEEE Trans. on Communications* **29**, 11–18.

33. Garey, M. R., and D. S. Johnson (1979). *Computers and Intractability: A Guide to the Theory of NP-Completeness*. Freeman, New York, NY.

34. Geman, S., and D. Geman (1984). Stochastic relaxation, Gibbs distributions, and the Bayesian restoration of images. *IEEE Trans. on Pattern Analysis and Machine Intelligence* **6**, 721–741.

35. Greening, D. R. (1990). Parallel simulated annealing techniques. *Physica D* **42**, 293–306.

36. Greening, D. R. (1991). Asynchronous parallel simulated annealing. In L. Nadel and D. L. Stein (Eds.), *1990 Lectures in Complex Systems*, 497–507. Addison-Wesley, Redwood City, CA.

37. Griffeath, D. (1976). Introduction to random fields. In J. G. Kennedy, J. L. Snell, and A. W. Knapp (Eds.), *Denumerable Markov Chains*, 425–458. Springer-Verlag, New York, NY.

38. Grimmett, G. R. (1973). A theorem about random fields. *Bull. of the London Mathematical Society* **5**, 81–84.

39. Grötschel, M., L. Lovász, and A. Schrijver (1981). The ellipsoid method and its consequences in combinatorial optimization. *Combinatorica* **1**, 169–197.

40. Hertz, J., A. Krogh, and R. G. Palmer (1991). *Introduction to the Theory of Neural Computation.* Addison-Wesley, Redwood City, CA.

41. Hinton, G. E., T. J. Sejnowski, and D. H. Ackley (1984). Boltzmann machines: constraint satisfaction networks that learn. Technical report CMU-CS-84-119, Computer Science Department, Carnegie-Mellon University, Pittsburgh, PA.

42. Hopfield, J. J. (1982). Neural networks and physical systems with emergent collective computational abilities. *Proc. of the National Academy of Sciences USA* **79**, 2554–2558.

43. Hrycej, T. (1990). Gibbs sampling in Bayesian networks. *Artificial Intelligence* **46**, 351–363.

44. Isham, V. (1981). An introduction to spatial point processes and Markov random fields. *International Statistical Review* **49**, 21–43.

45. Kagno, I. N. (1946). Linear graphs of degree ≤ 6 and their groups. *American J. of Mathematics* **68**, 505–520.

46. Karp, R. M. (1972). Reducibility among combinatorial problems. In R. E. Miller and J. W. Thatcher (Eds.), *Complexity of Computer Computations*, 85–103. Plenum Press, New York, NY.

47. Kinderman, R., and J. L. Snell (1980). *Markov Random Fields and their Applications.* American Mathematical Society, Providence, RI.

48. Kirkpatrick, S., C. D. Gelatt, and M. P. Vecchi (1983). Optimization by simulated annealing. *Science* **220**, 671–680.

49. Malka, Y., and S. Rajsbaum (1992). Analysis of distributed algorithms based on recurrence relations (preliminary version). In *Proc. of the (1991) International Workshop on Distributed Algorithms*, 242–253. Springer-Verlag, New York, NY.

50. Malka, Y., S. Moran, and S. Zaks (1993). A lower bound on the period length of a distributed scheduler. *Algorithmica* **10**, 383–398.

51. McKay, B. D. (1998). Isomorph-free exhaustive generation. *J. of Algorithms* **26**, 306–324.

52. Moussouris, J. (1974). Gibbs and Markov random systems with constraints. *J. of Statistical Physics* **10**, 11–33.

53. Pearl, J. (1986). Fusion, propagation and structuring in belief networks. *Artificial Intelligence* **29**, 241–288.

54. Pearl, J. (1987). Evidential reasoning using stochastic simulation of causal models. *Artificial Intelligence* **32**, 245–257.

55. Pearl, J. (1988). *Probabilistic Reasoning in Intelligent Systems: Networks of Plausible Inference.* Morgan Kaufmann, San Mateo, CA.
56. Prisner, E. (1995). *Graph Dynamics.* Longman, New York, NY.
57. Rajsbaum, S., and M. Sidi (1994). On the performance of synchronized programs in distributed networks with random processing times and transmission delays. *IEEE Trans. on Parallel and Distributed Systems* **5**, 939–950.
58. Scheinerman, E. R., and D. H. Ullman (1997). *Fractional Graph Theory: A Rational Approach to the Theory of Graphs.* Wiley, New York, NY.
59. Schuster, H. G. (1995). *Deterministic Chaos: An Introduction.* VCH, Weinheim, Germany.
60. Spitzer, F. (1971). Markov random fields and Gibbs ensembles. *American Mathematical Monthly* **78**, 142–154.
61. Squire, M. B. (1998). Generating the acyclic orientations of a graph. *J. of Algorithms* **26**, 275–290.
62. Stahl, S. (1976). n-tuple colorings and associated graphs. *J. of Combinatorial Theory, Series B* **20**, 185–203.
63. Stanley, R. P. (1973). Acyclic orientations of graphs. *Discrete Mathematics* **5**, 171–178.
64. Trivedi, K. S. (1982). *Probability & Statistics with Reliability, Queuing, and Computer Science Applications.* Prentice-Hall, Englewood Cliffs, NJ.
65. Vo, K.-P. (1987). Graph colorings and acyclic orientations. *Linear and Multilinear Algebra* **22**, 161–170.
66. Wuensche, A., and M. Lesser (1992). *The Global Dynamics of Cellular Automata: An Atlas of Basin of Attraction Fields of One-Dimensional Cellular Automata.* Addison-Wesley, Reading, MA.

Index